The River Cottage

Chicken & Eggs Handbook

The River Cottage Chicken & Eggs Handbook

by Mark Diacono

with an introduction by
Hugh Fearnley-Whittingstall

www.rivercottage.net

BLOOMSBURY
LONDON · NEW DELHI · NEW YORK · SYDNEY

For my sister, Rachel

First published in Great Britain 2013

Text © 2013 by Mark Diacono
Photography © 2013 by Mark Diacono
Illustrations © 2013 by Toby Atkins

Bloomsbury Publishing Plc, 50 Bedford Square, London WC1B 3DP
Bloomsbury Publishing, London, New Delhi, New York and Sydney
www.bloomsbury.com

A CIP catalogue record for this book is available from the British Library

ISBN 978 1 4088 1791 9
10 9 8 7 6 5 4 3 2 1

Project editor: Janet Illsley
Designer: Will Webb
Illustrator: Toby Atkins
Indexer: Hilary Bird

Printed and bound in Italy by Graphicom

www.bloomsbury.com/rivercottage
www.rivercottage.net

Contents

If you've spent any time watching chickens ranging free in a friend's yard or garden, beady-eyed and bouffant as they cluck and fuss, peck and preen – or just run away from you – it might not be immediately obvious that they are highly political creatures. Okay, it's unlikely you'll ever see them in slogan T-shirts and dubious woolly hats, bearing placards and marching on Westminster – chickens have always had trouble with organised group protest, so we've had to do it for them. And done it we have, because these endearing birds, and the eggs they lay, are at the cutting edge of the current debate about food.

I can't think of any other farm animal (although the pig comes close) that so comprehensively encompasses all the tricky issues that relate to what and how we eat in the developed world now. Animal welfare, intensive farming, organics, food wastage, our attitude to meat-eating, the power of multinational conglomerates, the power of the consumer to change all of the above – it's all here, in an eggshell.

I'm not suggesting, of course, that this is principally intended as some kind of polemic. It's not. It's a confidence-inspiring, witty and beautifully written handbook that will tell you all you need to know about keeping your own chooks. Nevertheless, the way I see it – beyond its practical use – it is fighting its own corner in the ever-challenging world of food politics. Because once you are looking after chickens, and have the wherewithal to produce eggs and/or meat for yourself, at home, you can take a meaningful step in life: a step away from anonymous, soulless, mass-produced, poor-quality food, and towards the satisfaction and fulfilment of a more self-reliant approach to feeding yourself and your family.

So as well as plying us with valuable and engaging information, I feel Mark is also making us aware that deciding to keep your own chickens is a significant act – a peaceful but potent protest against our food production system as it currently stands. I don't have the stats to prove it, but I'd confidently bet that chicken keepers are more than averagely likely to source the rest of their food closer to home, and higher up the welfare ladder. I sense this because it's clear that chicken keeping is no longer the sole preserve of rural-dwelling smallholders. Increasingly there are suburban, urban and, who knows, maybe even a few high-rise chicken keepers. Most of us know somebody who's taken the poultry-rearing plunge.

Enthusiasm, an enlightened attitude to food and a healthy appetite for good things may inspire you to take up chicken keeping, but they won't get you through the enterprise unaided. Taking charge of a living creature is always something to be done with care and forethought, and it would not be responsible to start keeping chickens without doing some research first. That's where this book comes in.

Mark's expertise, delivered in his warm and accessible style, will fully inform you about choosing, keeping and caring for chickens, whether for eggs or meat, or both. From choosing a breed and constructing a henhouse to slaughtering your own birds and preparing them for the kitchen, everything is covered.

When I first started keeping chickens at the original River Cottage over 15 years ago, I was amazed at how much pleasure it gave me. I immediately felt empowered and more connected to the land around me. I also had the deep satisfaction that comes from knowing your birds are contented and able to express their full range of instinctive natural behaviours. Good husbandry and free living is fundamental to the feel-good factor of raising your own livestock, on however small a scale: they feel good, so you feel good about eating them. Add to that the knowledge that the eggs and meat are as good as it gets – not just free of the chemicals and antibiotics that are often routine in intensive systems, but also higher in certain nutrients (omega 3s, for example) than the standard products of an industrial approach.

It's an overwhelming case, isn't it? And it's fun too. These birds are fascinating to observe and easy to look after – and it's hard not to be fond of any creature that produces a fabulous fresh egg for your breakfast on a fairly regular basis. I've had a flock for almost two decades now and I really couldn't imagine life without them.

And I know Mark feels the same way. I've known him for a decade now, and I'm constantly discovering (with only the mildest envy and irritation) that his talents and experience extend far beyond the amazing work he did during his period as head gardener and course tutor at River Cottage. He has, of course, two previous River Cottage handbooks to his name (*Veg Patch* and *Fruit*), which are as enjoyable to read as they are authoritative. With this third volume, he completes a kind of earthy triptych on the joys of a more self-sufficient life. He also runs his own highly successful organic smallholding, Otter Farm, not far from me in East Devon. So, whether you intend to launch yourself wholeheartedly into the adventure of raising your own food, or merely dabble in the shallows (you'll get fully hooked in the end, believe me), there really couldn't be anyone better placed or better qualified, nor more passionate to be your guide, than Mark.

Even if you're not yet sure if you want to commit to full-time chicken keeping, this book is worth having for the glorious recipe section alone. It's full of the kind of mouth-watering, soul-warming chicken and egg dishes that we find irresistible – the ultimate roast chicken, sumptuous pâté, silky custard, billowy meringues, to name but a tempting few. It's a perfect illustration of how the fairly minimal, laid-back work of chicken keeping can pay dividends in the kitchen.

In short, if buying your first couple of Columbian Blacktails or Orpingtons is a milestone on the road to a better understanding and appreciation of the food you eat – and I unequivocally think it is – then this handbook is the signpost that will set you off in the right direction, with a bounce in your step, from the very beginning. I know you'll enjoy the journey.

Hugh Fearnley-Whittingstall, East Devon, March 2013

Preparing to Keep
Chickens

ce Range
Eggs

There is a shade of yellow you may not be familiar with. Almost orange, it's unlikely to be a colour you would shortlist for your bedroom walls or your wedding attire, but once you've seen it at the breakfast table nothing else will do – it's the colour of the yolk of a home-produced egg.

It is this warm yellow, and the flavour that comes with it, that convinces many of us to keep chickens. That first egg from your own chickens, boiled – for some reason it must be boiled – is really something special. Once in a winter blue moon, when laying drops off and I'm forced into buying eggs from a shop, I am reminded of the difference between even the very best organic free-range eggs and those lifted, still warm, from the nest box: they are like chalk and chestnuts.

Aside from the eggs and meat they provide, chickens are a real pleasure to have around. Faintly bizarre, alert, often highly strung and naturally inquisitive birds, they are compelling to watch. Their day is a jumble of foraging, nesting, preening and picking, dust bathing, perching and playing.

Taking time to observe them is very much part of what makes home-raised eggs and chicken essentially 'Slow Food' and renders the exchange of a happy, nurtured life for eggs and meat an altogether fairer one. It makes food part of what you do, rather than what you eat.

If you have a veg patch or allotment, chickens come into their own. We allow ours to strip graze over the beds in winter, clearing away weeds and any bedraggled leaves, picking over the soil and feeding on potential pests such as leatherjackets, fertilising the soil as they go.

Keeping chickens takes a little preparation and some investment to get started but after that it is largely straightforward. Provide them with a house, enough space to express natural behaviours, give them access to food, water and keep predators at bay and you'll have happy, productive chickens.

Many people take the next step and raise chickens to eat. Again, it is surprisingly simple to do provided you know the fundamentals, but it does bring with it the prospect of a predetermined, as opposed to a natural, end. Serious undertaking as this is, I believe it should be embraced. If you eat meat (or dairy, where many of the newborn males are slaughtered as they offer no milk) animals die as a result. Being straight with yourself about that and choosing how and under what conditions those lives and deaths occur is really what River Cottage is all about.

Keeping a few chickens takes you across the threshold from the veg patch into smallholding. It is one of the most rewarding things you can do when it comes to food. The eggs are sensational, the meat unparalleled, and in providing yourself with even some of the eggs and chicken that you eat, you're doing all kinds of good.

Nine out of ten chickens that are raised to eat in the UK live in intensive conditions, with no natural light, no freedom to exhibit natural behaviours, and no chance to pick through the grass for greenery and small creatures. In almost all senses, they are prevented from being chickens. Without space they expend little energy and they can be fed constantly – growing at such a rate as to go from egg to Sunday roast in just 40 days. It makes it possible for a chicken to be hatched, raised, fed, watered, housed, treated, killed, dressed, packaged, transported and sold for around the price of a pint of beer.

The price for our cheap meat is a life without pleasure and likely as not ailments that include ammonia burns, leg disorders and sometimes lung or heart failure. It is, by most definitions of the word, inhumane. Around 850 million chickens are reared to eat in the UK each year, the vast majority in unsatisfactory conditions.

Laying hens do not have it better. Lay this open book flat on the table: the area is, scandalously, about the same as the floor space an intensively reared laying bird has to live in.

There is, however, some cause for optimism. The Chicken Out campaign has been instrumental in raising awareness of the issues and in influencing more of us to buy higher-welfare chicken. Your pound has power. Buying free-range or organic eggs and meat supports those using non-intensive systems and encourages other producers to follow suit.

Raising a few chickens of your own often has an even more positive effect on animal welfare. For many, it seems to tear away the curtain that hides where our food comes from. It's not a difficult leap to make to imagine the madness of keeping your hens constantly indoors, in a space less than the size of a sheet of A4, and to want to pay a little more when you are buying chicken and eggs, to ensure other animals can enjoy some of the pleasures that yours do. The same applies to other meat and poultry – even if the higher price means eating it a little less often. I've included advice on what to look for when it comes to buying eggs and chicken, as you will undoubtedly need to do from time to time, even if you keep a sizeable flock of chickens.

Enjoying your chicken and eggs in the kitchen and at the table is, of course, equally important and I hope you'll find plenty of inspiration here. Although little comes close to the flavour of a perfectly roasted well-reared chicken, there are so many brilliant ways to flavour and cook a chicken – whole or jointed. To help you get the most value from the whole carcass, I've included a number of recipes that use the less obvious, yet tasty parts of the bird, such as the liver, and several delicious soups and salads that feature chicken as a contributor rather than the main player. If you're not already, I hope you'll soon be making your own stock, and have the confidence to joint a whole chicken.

As with growing your own vegetables, fruit and herbs, the pleasure goes beyond the flavour. Home-raised chickens and eggs look and taste different from those you'll find in the shops because of the life they are allowed to lead, and that feels as good to know as the food is to eat.

Buff Plymouth Rocks in a run

Things to consider before you start

If you are thinking of keeping chickens, either for eggs or to eat, let me encourage you to make the leap. It is easy, hugely pleasurable and the food is exceptional. For all the simplicity though, you will be dealing with live animals: care is essential and preparation is vital. There are a few questions you should ask yourself before you embark on the journey, some set-up costs to be aware of, and decisions to make before you even get close to bringing chickens into your life.

Spend an hour or two with chickens before you get some of your own. Handle a bird, get used to it and develop a little confidence with chickens while you are still able to hand them back to their owners. Friends who keep chickens may give you the opportunity and be willing to show you how; otherwise, ask potential suppliers. It is quite understandable to feel nervous of handling a live chicken for the first few times but with the skill and confidence that come from doing it properly you'll be able to enjoy and care for your birds to the full.

Eggs or meat?

Ask yourself why you want chickens: as pets, for a little movement around the garden, for the pleasure of their endearing chatter, yes, but do you want their eggs, their meat or both? Decide on this before you take the leap, as varieties are generally better suited either to laying, or being raised for meat. In some cases, they are good for both.

For eggs, you need only have hens; for meat birds, you can raise hens, cockerels or some of each. Flocks of dual-purpose birds are usually mostly hens, to ensure enough eggs. Layers usually live for 5–10 years whereas meat birds and dual-purpose chickens are slaughtered at their peak for eating, often at only a few months old.

If you are in any doubt, I'd suggest keeping a few layers first – to get used to having them around and the skills involved without having to deal with the prospect of their death so soon after starting.

How many chickens to keep

Chickens are social animals: keep two as a minimum, although you're likely to want more once you've taken the plunge. How many is more? The same rule of thumb applies as when taking on a veg patch: it's better to build up from a small success than back from a large failure.

Starting small allows you to gain a little experience without getting overwhelmed. You can focus on the chickens' wellbeing and get used to the responsibilities and pleasures of keeping them with a manageable group. You may well find that the number you start with suits you in the long term. If not, you'll move on to a larger flock with confidence and experience under your belt. If you intend to keep hens for eggs, bear in mind that a hen from a good breed of layers will lay around six eggs a week during the sunny months but will do so only for a year or two before numbers decline, while dual-purpose (i.e. breeds that are good for eggs and meat) or less prolific breeds – often pure breeds – will give you about half that number but may well lay through the winter and for more years.

A dozen or so chickens is the usual minimum size to raise as a group for eating (see p.80). To a large degree, the scale that you reach above this is determined by a combination of how many chickens you and your family and friends want to eat in relation to how long they take to reach dispatch time and the space you have available to rear them. If the idea of raising a group of birds for eating sounds intimidating, you can always raise just a couple of birds for meat – perhaps cockerels of a dual-purpose breed you've raised from eggs or chicks – and dispatch them yourself.

Starting small also allows you to invest in stages. Housing, other equipment and feed can add up – you only want to get as much as you need.

Time

Chickens will take up as much of your time as you allow – there's something quietly fascinating about their busy pecking and their social interactions that can steal the hours away, but the minimum time they need from you is much shorter. Allow 15 minutes at each end of the day to feed them, let them out, lock them in, check for eggs and to see that everything is okay. Automatic doors, feeders and drinkers may make life convenient, but even if you use them you'll still need to take time twice a day to check that your chickens are well.

On top of that, there's weekly cleaning. Manure and bedding should be removed and dirty areas cleaned. Whether this takes 30 minutes or longer depends on the number and nature of your chickens. Every couple of months allow for a deeper, more intensive clean. Most of these regular tasks vary little in length whether you have a few or a couple of dozen chickens.

The timing of your visits is also crucial. Let your chickens out as close to when the sun is fully up as you can, as they need light to lay and be at their happiest; lock them up as soon as possible after they have found their way into their shelter to minimise risk from predators.

Space and location

When you read about how much (or rather how little) space chickens need, it seems almost cruel. Three chickens per square metre of floor space in their shelter, plus the same outside, seems pitifully small, but it will do. Give them more room to move about and they'll appreciate it. Add to that space for the feed and water containers, as well as to store feed.

Layers with access to free range

Make sure your chickens have access to vegetation. It is an important component of a healthy diet, provides for their behavioural need to forage for some of their food and, given their jungle floor origins, is likely to add to their psychological wellbeing.

If you are keeping chickens for meat, bear in mind they can grow quickly and their feed requirements escalate. You're likely to need a larger area than you would for the same number of layers, and you may well need space to move them to if the grass wears thin.

Chickens and their housing are valuable and security is a consideration – avoid keeping them in your front garden if it is easily accessible to those passing.

Chickens and children

Generally speaking, chickens and children go together very well. Chickens rarely peck the hand that feeds them, although a little enthusiasm to get at their dinner can lead to a light tap once in a while. Once your child is aware that the chicken needs looking after rather than chasing, and that they can play as much of a part in that as an adult, they are usually fascinated.

An appropriate choice of breed can go a long way to forging a good bond between child and chicken. Some are naturally calmer than others – hybrids more often than not. See pp.21–5 for more information about breeds.

A young chick being transferred to the chicken house

The relationship between child and hen will always be best where there is ongoing contact that starts as early as possible. Show your children how to feed your chickens and encourage them to handle hens once in a while to develop confidence around them. The feeling is usually reciprocated.

Keeping chickens is a fine way of illuminating all that can be good in human relationships with the animals that provide us with food. With your help, children often quickly appreciate the importance of providing animals with the conditions they thrive in and enjoy their eggs and/or meat in return.

Cockerels can be quite a different story. Some seem to be aggressive from the moment they reach adulthood, others can begin attacking almost overnight, often picking on children who they see as a good size to dominate. For more on dealing with cockerels, see pp.126–9.

One final point: it may seem a little fussy, but it is wise always to wash your hands after handling chickens and their eggs. This is especially important for children and pregnant women. Chickens can pass on disease and harmful bacteria (including via their eggs). A good wash with soap and hot water prevents the risk of this happening.

Chickens and household pets

Adult chickens and cats quickly form their own hierarchy – i.e. the chicken bosses the cat should it be dim enough to get too close. There is more room for danger if you are rearing young chicks and/or bantams. Bantams (see p.25) are almost always bold enough to look after themselves but cat attacks do occur. Chicks need to be kept enclosed, with a roof too, to keep any neighbourhood cats at bay.

Dogs and chickens can get along famously but this is not always the case. To promote domestic harmony, introduce them to each other carefully and in instalments, and with the dog on a lead. Although there may be some uncertainty and/or noisy curiosity to begin with, they usually get accustomed to each other surprisingly quickly. Be aware though – dogs have a prey reflex and the sight of a chicken darting at speed can trigger it. Even a playful chase can cause serious harm or worse to your chicken, so be mindful and never leave a dog alone with your chickens.

That said, having a dog shouldn't put you off keeping chickens. With a little care and attention they will coexist perfectly happily, with some dogs even herding the chickens into their house at night, but the onus is on you to minimise the opportunities for misfortune.

Rabbits and chickens are good companions. You can allow them to share a run but they should be housed separately. Be aware of any burrowing undermining the security of the enclosure though – you don't want to allow your rabbits out or predators in.

A Wyandotte cockerel making its presence known

Neighbours

Neighbours may be less fond of your chickens than you are, so site them in the most considerate location you can. Keeping neighbours on side will go most of the way to avoiding disputes which can easily escalate when noise, sights and smells are involved.

Be neat; keep everything to do with your chickens clean and dispose of bedding and manure regularly and out of sight/smell of neighbours.

If you intend to dispatch chickens at home (see pp.160–6), do so considerately, out of sight, keeping all signs of blood, feathers and the dead chicken itself hidden. If you have close neighbours, it is well worth considering sending your birds off to be dispatched – the waste attracts pests as well as flies and needs disposing of immediately if you want to avoid unpleasant smells.

Consider whether you really need a cockerel. A daily dawn squawking is among the most effective of relationship sourers. I must admit to enjoying the sound but many find it deeply irritating and noise pollution is something that is taken seriously by most local authorities, so your angry neighbours may well find a receptive ear.

Remember that it is your responsibility to keep your chickens in, not your neighbours' to keep them out. Even a single chicken can do considerable damage to a veg patch or garden border in little time and their poo isn't something you want to feel squelch between your toes as you walk across the lawn. And if it's a cockerel that finds its way into someone else's garden, it may be aggressive to people and pets. So take care of the fencing and keep your birds in your garden.

Be positive about your chickens: let everyone know ahead that you are getting them to soften any potential for upset, offer a box of eggs once in a while, and ask if your neighbours' children would like to feed the chickens.

Costs

As with most hobbies and pastimes, start-up costs can be considerable, but there are ways to keep expenditure to a minimum.

- **Housing** can be a substantial start-up cost. If you have materials to hand and a creative, practical disposition you may be able to knock a house together fairly cheaply, or you can spend hundreds of pounds on a shelter with a run – the choice is yours. Research ahead so that it doesn't take you by surprise (for more on housing, see pp.40–5). You'd be amazed how many people turn up at a supplier to buy some chickens without having organised their housing.

- **Chickens** usually come at a cost. A good laying hen may set you back the same as a reasonable bottle of wine and chicks considerably less than a glass, but if you choose rare breeds that are in demand the price can quickly escalate. If you are less particular about the breed and age, it is possible to source your chickens almost for free: the British Hen Welfare Trust (see Directory, p.246) may be able to provide you with ex-battery chickens for a nominal fee.

- **Feeders and drinkers** are required and vary considerably in price. Again, research beforehand. See Directory (p.246) for recommended suppliers.

- **Feed** is your primary ongoing cost. The amount of feed your chickens will need depends on how free-ranging they are, whether they are pure breeds or hybrids (hybrids often eat more), and whether they are being raised for eggs or meat. As a guide, estimate 125g of feed per adult chicken a day. Try not to skimp on quality. A good GM-free, ideally organic feed will pay dividends for your chickens' health and the quality of the eggs and meat you enjoy. For more detailed information on feed see pp.113–19.

- **A vet** will charge for their services. The overall cost is hard to predict, but be aware that illness, disease and injury, rare though they will hopefully be, must be treated.

- **An incubator and/or a brooder** will be needed if you intend to raise birds from eggs or day-old chicks. Both items can be expensive, although you can easily make a brooder. See pp.51–4 for more on incubators and brooders.

If you can, buy everything in person. This minimises transport costs, you get to see and examine what you are buying, and you have the opportunity to build up a relationship with the supplier.

Cockerels or not

The non-laying, often cantankerous, occasionally aggressive, persistently early-rising cockerel provides at least one useful counterbalance via his sperm. If you want to raise chicks from your own eggs, keeping a cockerel is the easiest way of doing it. But you don't need a cockerel to produce eggs – the hens do that very happily by themselves – and many domestic hen keepers do without one. Decide considerately, bearing in mind that a cockerel's crowing may annoy neighbours.

Although cockerels can be perfectly mild-mannered and content to keep their place at the head of the chicken hierarchy without feeling the need to challenge their owners, it is not always the case. For more about dealing with aggressive cockerels, see pp.126–9.

Legalities

If you own your house, check the deeds for any covenants preventing you from keeping chickens in the grounds of your home; if you are renting, speak with the landlord or local authority. It may be that no one will enforce a covenant even if one is in place, but speak with your neighbours before getting chickens to see whether they have any objections.

If you are considering erecting a building in which to house your chickens, seek advice from the local authority, as well as the Environment Agency if you are near a waterway.

The legalities for those with less than 50 hens are relatively straightforward: you may sell unwashed, ungraded eggs at the 'farm' gate or locally door to door; you can also sell eggs at a local market as long as you provide your name and address, advise customers to keep the eggs chilled and suggest a 'best before' date no more than 28 days after they were laid. See p.72 for more information if you are planning to sell eggs.

Registering with a vet

Most chicken ailments are easily identifiable and treatable but specialist advice should be easy to come by if you ever need it. A stray dog, an aggressive infection or an unidentifiable disease may all be beyond the treatment of the enthusiastic amateur. Also, you may not want to be the one who draws your chicken's life to a close if intervention is required – it is so much more bearable if you are already on the books of a local vet.

Research and find a vet ahead of trouble. Some that specialise in pets rather than livestock do not treat chickens, so you cannot simply assume that your local veterinary practice is at hand to help when needed.

Understanding breeds

Since humans began domesticating birds, they have been selected and bred to emphasise desirable characteristics such as speed of growth, laying rate and aesthetics. We are still doing it today.

Each breed of chicken has its own particular traits of appearance, behaviour, flavour and resistance to disease. It's not essential to familiarise yourself with the fine details of their genetics but do get to know some of the common breeds. Bear in mind that within every breed a degree of variation exists, especially when it comes to colour. And of course, males and females of the same breed can be markedly different in plumage colour, comb size, tail feather structure and other physical attributes.

Breed types

The combination of breed type and breed category (see right) indicates much about the character and likely purpose of the bird. The breed type – purebred, hybrid, strain or mixed breed – tells you about the bird's origins, i.e. whether it is a cross, and if so a whether a complex or simple one, which can help in identifying a breed that suits your purpose. More information about particular breeds best suited for laying or rearing for meat, or a combination of both, is provided in Chickens for Eggs (see pp.56–73) and Chickens for Meat (see pp.74–83).

Purebreds These chickens are true to their type. Having been bred with chickens of the same breed for a few generations, they display all the main characteristics of that breed. As such, breeding a purebred with another of the same breed should result in offspring very similar to the parents.

Purebreds are a good option when you know the characteristics you want and if you want to breed chickens true to a specific breed. They are also popular with people who want to show their chickens competitively.

Another characteristic many purebreds have is that they lay steadily (rather than prolifically) but do so over a number of years.

There are a couple of hundred chicken breeds, half of which are reasonably common, largely thanks to small-scale enthusiasts.

Hybrids A hybrid results from crossing two purebreds, i.e. the father is of one breed and the mother a different breed. The aim of crossing breeds is usually to produce a chicken that has the most desirable traits of both parents. They tend to have been produced with a specific goal in mind: producing fine tender meat in a short period, being prolific and regular layers, etc, which may suit your requirements. Many hybrids have also been bred to be healthier than purebreds.

Hybrids are not the ideal choice if you want to breed your own chickens as the characteristics of the offspring are unpredictable, being potentially like, or indeed entirely unlike, either parent.

Unlike most pure breeds, many hybrids lay prolifically but only for a year or two, before laying rates decline considerably.

Strains These are variations on a purebred or hybrid. They have all the main characteristics of the breed but by careful breeding, particular characteristics such as a specific colouring or skin colour are accentuated to produce a strain. These are often available from only one breeder.

Mixed breeds These chickens have uncertain origins, usually the result of at least one generation of hybrids producing offspring, but often an imprecise combination of many breeds. They are a popular choice for the home chicken owner as, having been bred intentionally to be so, they are usually resilient, reliable chickens that lay reasonably well. The downside of mixed breed chickens is that it is hard to replicate any desirable traits they exhibit, and if they interbreed randomly (rather than being selected to replicate desirable characteristics) they often produce smaller, less productive, sometimes less healthy offspring.

Breed categories

There are a number of categories into which chickens are grouped, based on the qualities and characteristics of their breed.

Egg layers The most prolific breeds in this category lay between 150 and 300 eggs in a year. As a rule, the more highly productive layers are usually hybrids and they lay at this impressive rate for a year or two before their productivity declines, often quite rapidly. Hybrid layers rarely become broody, which is an advantage if you want birds for eggs.

If, however, you want to breed chickens you may want to consider a purebred layer, many of which are pretty good sitters. Purebred layers typically produce fewer eggs per year than hybrids but do so over a longer period. While laying breeds are edible, they tend to be without the generous covering of meat that makes for good table birds.

A white bantam with an Orpington hen

Meat birds Most birds in this category are hybrids, bred to have larger breasts, a bigger frame and a fast rate of growth. Although hens of meat bird breeds will lay perfectly delicious eggs, they are usually not prolific layers.

Dual-purpose birds For smallholders, these breeds may be the holy grail as they are pretty good layers, yet grow well enough for some birds (often the non-laying cockerels) to be grown for meat. Dual-purpose breeds also tend to be calmer than most others.

Show birds There are some breeds, such as Cochins and Old English game birds, which are kept largely for their looks and the pleasure of putting them into show competition. They are often poor layers and rarely make good eating but keeping them for showing helps maintain the genetic diversity of chickens.

Bantams These are small chickens – either miniaturised versions of existing breeds or naturally occurring small breeds, such as the Japanese bantam. Although bantams don't make good meat birds and rarely lay eggs in huge numbers, their eggs are perfectly good to eat and they make an excellent choice for those with limited space. Bantam breeds that are smaller versions of a particular breed have the same colour eggs as the non-bantam birds.

Rare breeds Although not strictly a category of breeds in the way the others are, it is worth considering rare breeds when planning for your flock. As with most of our food, from vegetables and fruit to livestock, commercial production focuses on a fairly narrow range of breeds and varieties. Disease resistance and predictable productivity, both very good characteristics, ensure reliability for the commercial food chain.

Rare and heritage breed suppliers, and those who buy from them, help to keep the range of chicken breeds wide and the genetic pool diverse. Rare breeds are often very individual and characteristic. They may lay less than the most prolific breeds or be relatively slow growers but, even if you keep just one or two within a flock of more familiar breeds, you'll not only be doing your bit for keeping rare breeds in existence by providing a market for those who are breeding them, you'll also have a distinctive flock. For more information about obtaining rare breeds, see Directory (p.246).

Understanding
Chickens

Whatever animal you bring into your life, if you get to know how it operates and communicates, and you understand its basic biology, it will help you to make that animal's life a happy one.

The chicken's body

Becoming familiar with the visible parts of the chicken will help you care for it, as you'll be able to identify where to look for most ailments and apply treatments should you need to. It also enables you to pick out similarities and differences in breeds, and helps you communicate more confidently with others – including those you may be buying from, or selling to, and the vet.

Comb This fleshy, usually red structure sits atop a chicken's head. Releasing heat through the comb is one of the ways a bird regulates its body temperature – large floppy combs usually originate from chickens bred in hot countries, and vice versa. Combs may differ from breed to breed but are typically larger in the male.

Eyes Chickens have small eyes, set on each side of the head. They blink from the bottom upwards. Eyes are usually coloured yellow and black, grey or red-brown.

Ears The ear openings, on either side of the head, are surrounded by a small patch of bare skin with a hanging red lobe, and are usually hidden by small feathers. The colour of the skin patch indicates the colour of the eggs the bird will lay: red and the eggs will almost certainly be brown; white and the eggs will be white too. Breeds that lay blue or green eggs, such as Araucana, have red patches of bare skin.

Beak A chicken uses its beak to feed and drink but also to pick through its feathers, grooming itself. Beaks are hard, usually yellow in colour, with the lower half sitting slightly behind the upper. Newborn chicks have small beaks, with a hard growth at the end, known as the egg tooth, which helps the chick break out of the shell when hatching.

Nostrils These are housed on either side of the upper half of the beak. They are usually visible but may be partly covered in some breeds by the lower part of the comb. Chickens have a fine sense of smell.

Wattles Under and on either side of the beak you'll find two fleshy lobes of skin known as the wattles. These are usually red. As with the comb, the size and form may vary with breed, but the male's tends to be larger.

1. Beak
2. Comb
3. Nostril
4. Wattle
5. Ear
6. Neck
7. Wing
8. Saddle
9. Tail
10. Vent
11. Breast

Neck Most chickens have a long, narrow neck, ideal for keeping a good field of view and allowing some reach up and over to get at vegetation. The small feathers that cover the neck are known as hackle feathers.

Breast This is the area at the front of the bird, between the neck and belly. It is larger in breeds kept for meat.

Saddle On the back of the bird, between the neck and the tail, lies the saddle (though it is inadvisable to use it as such). The wings join the body on either side of the saddle.

Wings The two wings, one on either side of the bird, enable the chicken to fly. By 'fly' I mean indelicately and momentarily relieve itself from the constraints of gravity's pull, but this may be enough to take it over a fence or hedge. If this is the case, consider clipping the wings (see p.129). The wings have three types of feather which slightly overlap each other: small rounded feathers known as coverts that are found closest to the body; primary feathers, the largest of the wing feathers, found at the end of the wing; and secondary feathers between the two.

Tail Both male and female have small tail feathers, in a fan or arching pattern, that aid balance – the male's are often hidden by the larger display feathers. This tail is often exaggerated in the cockerel by long, fanned feathers, often colourful but serving no purpose other than decoration. The few tail feathers that may arch above the rest in the cockerel are known as sickle feathers.

Legs and feet Almost all chickens have four toes: three facing forward and one back. Each toe has a nail. Chickens also have a bony toe-like growth called the spur, which pokes out from the inner part of the leg. In the male this spur can become very pronounced and sharp and is one of its primary weapons – so beware a crabby cockerel.

The legs are covered in overlapping scales, usually yellow in colour, that protect the lower part of the leg. While almost all breeds have small feathers on the thigh, some breeds – known as booted breeds – have feathers extending down the leg and even over the toes.

Skin A chicken's skin is delicate and is easily damaged when it is alive and equally so when it is being plucked (see p.170). The colour of the skin is almost always somewhere on the spectrum between white and yellow and, to a small degree at least, a diet that is high in greens and corn can move it further along the line from white towards yellow.

Vent This is the opening at the rear of the bird, beneath the tail, where the chicken releases waste material as faeces. As with all birds, chickens don't urinate, releasing all their waste when they poo.

Feathers A chicken's feathers take a variety of forms depending on function and which part of the body they cover. Most are lost and replaced during the autumn moulting period.

A chicken's feathers are grouped into two main types. The large, sturdy feathers that form the wings, tail and main upper body of the bird are known as contour feathers, while the softer feathers, close to the bird's body, are without the central spine of the contour feathers and are known as down feathers. These may be given additional names that relate to the part of the body where they are and/or the function they carry out (see Neck and Wings, above, for example).

Chickens moult (lose their feathers) each year, usually during autumn when light levels drop off and temperatures dip. All may fall out almost simultaneously or, more usually, they steadily lose their feathers and replace them over a period of a few weeks. Hens almost always stop laying during the moult, as replacing their feathers is tiring and stressful, but most begin again as the days lengthen in spring. Cockerels undergo a less drastic moult, and are usually not so exhausted by it, though they may become temporarily infertile until fully feathered again.

A moulting Orpington hen

The hierarchy

Chickens are social creatures, building and managing a hierarchy within their flock. If there is a cockerel, he is the head, but whether one is present or not there is still a ladder of ranking amongst the hens. This is the famous pecking order. It is usually ongoing, with the hierarchy altering here and there, rather than being fixed in stone. If new birds are added, this can cause quite a kerfuffle as the flock busy about establishing a new pecking order.

The cockerel rules the roost, keeping order, resolving disputes between his hens and playing his role in establishing the hierarchy of the flock. In return he oversees the eating order and mates when he pleases. If more than one cockerel is present, you may find that sub-flocks evolve – each cockerel with his own harem of hens. Occasionally scraps may still take place between the cockerels. This is quite normal, but do watch out for anything more serious and be prepared to separate the birds if need be.

Communication

Listen out for your chickens' language. You should be able to identify quite a range of different noises, from the loud and demonstrative to the seemingly insignificant. They all have a purpose, even if simply to remind each other of their presence within the flock. Often a noise can mean something very particular.

During the first few weeks that we had hens, we couldn't understand their odd, persistent, loud cackling noises, usually in the morning. We thought it might be distress of some kind until we realised it coincided with one or more of the hens laying an egg. The layer started the cackling and, more often than not, the other hens joined in.

The familiar sound of clucking hens seems to be just idle communication. Equally famously, the cockerel's crow is all about territory: he is saying loud and proud that this is his patch. Most cockerels crow at any time of day or night, but often make the most fuss of doing it first thing, as the sun comes up and they awake to a new day.

Squawking and loud squeals are usually signs of alarm. It can be in response to being surprised or picked up, or it can signal the presence of a predator. You may also hear a growl from a hen, especially if you lift a broody one from her eggs – it is a warning sign and may be followed by a swift peck.

Listen, too, for soft come-and-join-me noises from the cockerel to the hens or the hens to the chicks, most usually heard at mealtimes when new food has been thrown into the enclosure.

Dust-bathing Silver Sussex hen

How chickens eat

Chickens eat quite differently from us. They have no teeth and so rely on tearing off pieces of food with their beak or eating grains and pellets whole. Following the path of the digestive system, the food passes through the oesophagus into an internal pouch, called the crop, where initial softening of the food takes place. Once the crop is full, the breast will look and feel slightly swollen.

Food moves from the crop to a small stomach called the proventriculus, where it continues to break down. From here, it moves on to the gizzard where muscular contractions cause small stones (that chickens ingest and hold in the gizzard) to grind the food up. Now in a suitable form to pass into the intestines, the food is extracted of all its nutritive value and the remainder excreted.

When the crop is full a chicken cannot eat any more, so it makes good sense to let your birds have their essential food first, before any treats. It's for this reason that we feed our hens their all-important layers' pellets first thing, and give them grain in the afternoon.

Natural behaviours

Encouraging your chickens to exhibit their natural behaviours – those patterns of life that allow it to live happily and express its innate chickeniness – does much to ensure that they will lead proper, happy lives.

These behaviours are simple, seemingly small things that together add up to a good quality of life for the chicken. They include:

- dust bathing
- preening
- foraging around for food
- the opportunity to socialise, communicating visually and vocally with other chickens
- playing

In order for your chickens to fully express these behaviours, you need to provide access to certain things. Each is covered in greater depth in the book, but you should have them at the front of your thoughts from the moment you consider keeping chickens.

Food and water Your chickens will need access to plentiful food of the type most suited to their age and breed, as well as an uninterrupted supply of water (see p.46).

Shelter To flourish, your chickens require protection from the extremes of weather and seasons. Chickens can suffer damaged combs and toes in cold weather, especially if given inadequate shelter from rain and winds. Temperatures over 32°C can be very harmful to chickens, especially the larger meat breeds. You will need to allow more room per bird in the house if you live in a hot area and ensure good ventilation in hot periods. A shady position can help keep off the worst of the heat as well as the cold. Their housing should be waterproof, windproof, free from damp, able to be ventilated/cooled when temperatures exceed 30°C and offer shade from the heat of the sun (see pp.40–6). Semi-shade is their natural environment.

Sleep Chickens sleep at night, the darkness bringing a deep slumber that makes them both rest well and highly vulnerable to predators. It also makes night the best time to catch or move a chicken. They usually sleep on a perch, an evolutionary remnant of having naturally roosted up high in trees. Chickens will usually roost in the same place each night once they are in the habit, which makes life easy for you, as they return to their house when the sun goes down.

Natural light Chickens flourish best in natural light. Even if your birds are kept indoors for part of the day, they should have access to natural light too. Light levels control the rate of egg laying. Chickens can live with artificial light and vitamin D supplements to make up for the lack of sunshine but we shouldn't ask them to.

Grass and free-ranging Providing chickens with the opportunity to forage for part of their diet not only helps keep it more balanced, it allows them to do what they were meant to do: feed themselves. Access to the earth also gives them the opportunity to dust bathe, which they seem to take great pleasure in; it also helps keep some parasites in check.

Being able to roam in an open space also provides the flock with the opportunity to interact or separate, helping to establish and maintain the pecking order.

Variety Like us, chickens can become a bit bored when faced with the same-old, same-old every day. Although they don't need wholesale change, once in a while they do love something different to eat (such as a cabbage or kale hung from a post), or something new to clamber over (a pile of logs, perhaps, or a couple of pallets lashed into an upturned V). It provides a stimulating change, a chance to play and another way for the flock to act out its pecking order.

Free-ranging Buff Plymouth Rock hens

What You Will Need

Providing chickens with food, plentiful clean
water, a dry and clean house, and a safe environment in which to move and socialise requires a certain amount of core equipment. If you are starting with eggs or chicks, there are a couple of extra items of essential equipment, plus a few peripherals you may find useful.

The basic equipment you will need:
- Housing
- Enclosure and runs
- Feeder and drinker
- Food and food storage
- Bedding
- Bird carrier

Additional equipment needed if raising chickens from chicks or eggs:
- An incubator
- A brooder and possibly one or more heat lamps

Housing

The chicken house is the place your chickens will sleep and lay. It is usually the largest initial expense when starting to keep chickens. Take time to consider your options: choose well and it will be an investment that delivers for many years; pick a dud and your hand will be back in your pocket before you know it. Checking the housing out in person is always wise – as ever, supporting a local business that has been recommended is usually the best way.

You have various options to choose from: a traditional wooden house with or without an enclosure, a plastic house and run, an adapted garden shed or even a converted barn. There are many combinations and variations on the market or you may choose to build one yourself (see Directory, p.246). Whichever way you go, think not just of the building your chickens will sleep in but of what their whole enclosure or free-ranging space needs to provide.

Some houses come complete with an enclosed space or run, others are just the house itself. If the former, ensure the enclosed space is sufficient for the number of birds you intend to keep (see below); if the latter, decide how you will enclose your birds. Even if you intend to allow them to roam freely, you'll need to be sure your fences and hedges are chicken-proof.

There are legal minimum stocking rates (the space each bird should be afforded), which are aimed at the commercial chicken business and really don't add up to keeping a happy and healthy domestic flock. Ideally your chickens should have a minimum of 0.3m^2 each of housing space and 0.5m^2 each of run area. If you are moving chicks from brooder to the outdoors, allow at least 100 x 50cm of run space for up to three chicks, with an extra 30cm^2 for each additional chick. Generally, the more outside space your chickens have, the happier and healthier they are likely to be.

Their housing and/or outdoor area must also provide shelter from wind and rain, and offer protection from extremes of temperature. A well-insulated house is essential and the outdoor area should offer some shade for your chickens in the warmest weather.

Things to consider when you're buying housing:

- **Decide whether you need just a house or some form of enclosure too.** Free-rangers may be fine with only a house as long as the garden boundaries are secure, otherwise you will need a run or fencing to keep your chickens in and predators out. Foxes and pet dogs are the most common dangers and a secure house is essential for protecting your birds at night, when they are most vulnerable.

- **Make sure the housing will provide sufficient space.** The housing must give your chickens enough room to exhibit natural behaviours, including flapping, pecking about and dust bathing. Too little space and stress, conflict, disease and a reduction in laying are likely to result. A good rule of thumb is to allow 1m^2 of floor space in the house per three adult birds or five bantams. It makes sense to allow enough room for a few more chickens than you intend to start with, in case you decide to expand the flock.

- **Check the construction materials used.** Most houses and enclosed runs are made of wood, although in recent years plastic houses have become increasingly popular. Whichever you choose, avoid the cheapest. Thin wood, poor-quality fixings and speedy construction will give you a short lifespan. As a rule, hardwood is more durable than softwood but more expensive. If you are considering buying softwood housing, check that it has been treated to extend its life.

- **Check the nest boxes and perch.** The housing must provide nest boxes in which to lay (unless you plan to keep only meat birds) and a perch for the birds to roost on. These should be sturdy and, ideally, with the perch higher than the nest boxes (though not directly over them) as a hen's instinct is to

Chicken house on stilts

perch higher than it lays. Make sure there is sufficient headroom above the perches – the breeds you intend to keep need to be able to sit tall without their heads touching the roof.

- **Check for ventilation.** There should be sufficient air holes and/or adjustable vents. Ammonia and moisture from faeces can build up in static air within enclosed spaces, causing discomfort to and potential respiratory disease in your chickens; a good air flow will prevent this. If ventilation is adjustable, so much the better. Your chickens should have a draught-free place to sleep.

- **Check for access and security.** Satisfy yourself that each part of the house (and run if buying one) is accessible. You'll be collecting eggs daily, cleaning the house weekly and occasionally you may want to catch one of your birds. Ensure that the doors are securable too.

- **Consider whether the housing needs to be portable.** If you think you may need to move the housing, two smaller, lighter houses could be better than one large one. Ask yourself whether a design with wheels is best for you.

- **Consider regular maintenance.** Chickens dislike muddy conditions, so you will need to provide a clean environment. Good drainage, the room to roam freely and/or the ability to rotate their grass area are vital. The inside of the house must be easy to clean, as you'll need to change the bedding/floor covering material regularly.

- **Consider long-term maintenance.** Bear in mind that wooden houses (and enclosures) will require a little maintenance once in a while, particularly the odd treatment of (chicken-friendly) preservative to extend its life. Your supplier should be able to advise you about the most appropriate treatment.

Once the main considerations of the chickens are taken into account, whether you choose something A-frame shaped, of mini-shed design or a dome is entirely up to you. Plastic houses – increasingly popular – have much going for them. Having smooth, non-absorbent sides, they tend to be easier to clean than wooden houses with their delightful roughness and nooks and crannies. This keeps the likelihood of pests such as red mite (which loves a crevice to hide in) to a minimum, as well as doing away with most maintenance of the house. Being plastic they are not liable to rot and consequently have a long lifespan. They also tend to be lightweight in comparison to wooden houses. These advantages may come with a higher initial cost, although over time they often work out to be very good value.

Enclosures and runs

Your chickens will need to be secured with a boundary between their territory and the rest of the world. There are three main ways to do this: using a run made of a fixed framework and chicken wire, with an enclosure of moveable (perhaps electric) fencing, or with fixed fencing.

A fixed framework

A run made of a wooden framework with chicken wire mesh walls and ceiling is a popular way of enclosing chickens kept in a garden. The stronger the mesh used, the more predator-proof the run will be: 2mm upwards is the ideal thickness. If you are constructing a permanent run, dig the sides of the run at least 40cm into the ground to keep animals from burrowing in. If you want the flexibility of being able to move the run, you will need to give it a skirt of at least 30cm, pegged to the ground, to keep predators from burrowing in.

You can do without the 'roof' if you are confident that overhead bird attacks on young chicks are unlikely, or if the sides are high and/or electrified.

Moveable fencing

We use electric fencing, powered by a 12-volt battery with a small solar panel to contain and deter. The chickens get a light nudge through their feathers, but a fox gets quite a zap and often is warned off for good by it. Certainly since the electric fencing has been in place, we haven't lost any chickens. A little thought is needed though – you can't allow the grass to grow up and short out the supply (effectively allowing the battery to discharge all its power). You should check the equipment frequently to ensure it is working and have a spare battery fully charged at all times to replace the other one when it is charging. The ticking of the energiser should let you know everything is working; if in doubt, lay a blade of grass on the wire and you should feel a faint pulse through your fingers.

Fixed fencing

Whether you allow your chickens to roam freely or keep them enclosed with fencing or in a run is a matter of circumstance and preference, but even if they are free-ranging, you'll need to ensure that the boundaries are secure. Without an enclosure of some kind your chickens are more vulnerable to predators, including pet dogs, and they can escape into neighbouring gardens and areas of your own garden where they may damage valued plants. In time, they may become confident enough to stray further from their shelter, putting them more at risk. Chicken wire, ideally buried around 20cm deep or secured to the ground and attached to a series of posts, creates a semi-permanent enclosure relatively cheaply.

Wooden chicken house and run

Plastic chicken house and run

Siting

Many houses and runs are designed to be easily moved around to give your birds access to fresh ground; others are less easy to relocate. Either way, bear in mind the general principles of chicken keeping. In particular, avoid damp areas and any site that is prone to flooding, keep your birds sheltered from strong winds and make sure they have some shade. Chickens are usually fine in snow, although some can be startled to find their green/brown world turned white overnight and may be reluctant to leave the house. Remember to keep their water and feed accessible when snow has fallen.

If you have a traditional wooden house I would strongly recommend raising it off the ground (around 30cm or so) with blocks or legs, as rats are very adept at burrowing under and into a house. This also reduces the likelihood of the house rotting in the damp.

Feeders and drinkers

A feeder should hold plenty of food in a central barrel, dispensing it into a tray for your chickens to peck at as and when they feel the need to. Plastic feeders are inexpensive but they are usually light and liable to blow over in gusts of wind; galvanised feeders cost more but are robust and usually have a well-fitting lid to keep rain off. Feeders can be fixed, freestanding or suspended from a beam. If possible, choose one with a sloping lid to prevent chickens from using it as a launch pad to get over fencing, and ideally with a brim that is wider than the feed tray to keep the feed dry.

By all means consider making your own feeder. It can take whatever form you like, as long as the food is accessible to your chickens, but do ensure there is no risk of the food getting wet when it rains.

The size of feeder is very much dependent on the number and breed of bird, but as a general rule of thumb, allow 500g of storage space per bird. Ask your supplier for advice about your breed and, if in doubt, err on the generous side.

The choice is quite similar for drinkers, with drip or, more commonly, barrel dispensers available. Plastic drinkers are a better option than galvanised ones if you're planning to add cider vinegar to the drinking water occasionally (see p.124), as vinegar corrodes metal.

Plastic drinkers tend to have a shorter lifespan than galvanised, becoming brittle in the sun over time. As with feeders, the size of drinker required varies with breed, size and number of birds, but as a general rule allow a capacity of 1 litre of water per bird. Whatever you do, never let it run dry: chickens can get by for a short time without food but time without water is very stressful for them.

Hanging feeder

Fixed feeder

Drip drinker

Barrel drinker

Food and food storage

While free-ranging can provide some of the dietary requirements, commercial feeds ensure that a chicken gets everything it needs, being formulated to provide the right amounts of protein, energy, minerals and vitamins for its stage of life.

There are commercial feeds for layers and meat birds that suit their differing requirements; for example, adult layers need more calcium to keep up a steady supply of quality eggs. The quality of the feed is important too – it is not just the volume of food that will keep your birds healthy, but its components.

As a very broad rule of thumb, each laying bird will consume around 125g of food a day, and meat birds often more, depending on breed, age and time of year. (For more details on different types of feed, see pp.113–16.) It makes sense to buy feed in bulk – 20kg or 25kg sacks tend to be much better value than smaller bags.

Get yourself a galvanised food bin in which to keep bags of food. I've found a regular domestic waste bin holds two 25kg bags, which is a perfect size for our small flock of layers. I keep it just outside their enclosure, out of their reach, yet right where it's needed. Ensure that the lid fits snugly – a strong wind can lift loose lids off and heavy rain can ruin the feed in a few minutes.

Don't be tempted by wood or plastic containers – rats and mice will find their way in very quickly. If you have a large flock, there are many options for larger feed stores, or you can just add more smaller bins. Whatever you do, don't be tempted to leave bags unprotected – even in a locked store, something with a taste for it will find its way in.

Bedding

Chickens do most of their pooing at night, and it is essential to place something on the floor of the house to catch it. Bedding absorbs the worst of the droppings, making cleaning easier and less frequent, and helping to soak up some of the intensity of the smell.

If you have access to a supply, shredded paper makes very effective and cheap bedding. It doesn't last as long as other options though, so expect to replace it relatively frequently.

Wood shavings are popular, as they are cheap, soft, provide good insulation in cold weather and are particularly effective in absorbing the ammonia smells that can be so powerful. Try to get shavings that have had the dust extracted from them as chickens are susceptible to respiratory discomfort from dust.

Hemp bedding is increasingly popular. It is both highly absorbent and durable, balancing its comparatively high cost with the lower frequency of replacing.

Shredded paper bedding

Transporting a cockerel in a metal carrier

Straw and hay are not great options. They get compacted quickly and have a tendency to hold water within their hollow structure, which can lead to mould and in turn to respiratory problems in your chickens.

Chicks require bedding specific to their needs – softwood shavings are excellent (see p.104).

Bird carriers

Carrying crates are essential for transporting adult birds. Many suppliers and pet shops sell them. They are inexpensive, have a robust structure with proper securing doors and are easily cleaned and stacked. Birds need room to turn around and sit down so don't overcrowd them.

Baby chicks can be transported in cardboard boxes, well ventilated with holes. If moving only a few chicks, don't give them too large a container. Look for a box that provides no more than twice the floor space they need, to encourage them to huddle together and keep warm.

Give your birds access to water if the journey is over an hour – a clip-on drinker is ideal. You will need to provide food if the journey is over 12 hours long.

Be aware of heat; birds are very susceptible to overheating and can easily die. Never leave chickens or chicks in a container in direct sunlight, in a closed car when the temperature is over 27°C, or uncovered and exposed to wind and sun in a trailer or similar. Give them some shade and keep them cool in hot weather.

Brooders

If you are planning to start a flock with eggs or chicks, or add to one without using a hen to act as the mother, you will need a brooder. A brooder is simply an enclosed area that keeps chicks in a confined space, protecting them from the outside world and allowing you to provide them with everything they need in one place. They are usually kept indoors where warmth and safety is more easily provided.

Once chicks have emerged from their shell (or have been delivered) and have turned from wet little things into dry, fluffed-up chicks, you have to provide them with an environment that gives them the warmth and shelter they require in the early weeks, where you can feed and water them, and the best way of doing that is to create or buy a brooder. If you buy a brooder it will almost certainly come with a heat source to keep the chicks at the right temperature.

For one or two chicks, 50cm² is an adequate starting floor area, but this should be expanded by 15cm² or so for each additional bird. The sides need to be at least

50cm high and ideally your brooder should be rectangular; this makes it easier to provide a warm end (using a heat lamp, see p.52) graduating to a cooler end, which allows the chicks to find the location where they are most comfortable.

There are many different brooder designs from which to choose, depending on your taste and the scale of operation. You can make your own (see p.52), which will serve perfectly well while you are starting out and if you're only planning on rearing small batches of chicks once in a while. Your local supplier may sell brooders, or you might need to look online, especially if you want to buy a large brooder (see Directory, p.246).

Making a brooder

If you are thinking of making your own brooder it has to tick all the boxes of a bought one: providing warmth and shelter and protection from predators, being comfortable for the chicks and having some form of temperature control – usually provided by a heat lamp (see below).

A wooden box of suitable dimensions (see p.51) is ideal. Use well-secured mesh over the top to keep out predators if the brooder is in an outbuilding or garage. Adapt whatever fits the bill – a water tank or plastic storage container works well and is safe from the risk of fire from your heat source.

Don't use any containers that have held chemicals, as residues can survive even after thorough cleaning. Generally, cardboard boxes are best avoided too, as they pose a potential fire risk. As a one-off a cardboard box is adequate, provided you make it large enough to keep the heat source away from the very edge of the box to minimise the fire risk, and site the brooder inside the house and away from predators, including pets. Don't reuse a cardboard brooder, as it is impossible to clean thoroughly.

You will need an overhead heat source to mimic the warmth the chicks would usually get from the mother – most bought brooders come complete with one; for homemade brooders, you will need a heat lamp. You have two options with bulbs: heat lamp bulbs or reflector bulbs.

Heat lamp bulbs are available as infrared or white light bulbs. Infrared bulbs work perfectly well but, because of the wavelength they operate at, the temperature doesn't register well on a thermometer even though it is felt by the chicks. You have to rely on observation (see p.103) to be confident that the temperature is right.

Reflector bulbs give off more light than infrared bulbs but the chicks don't seem to mind and the light is directed downwards, leaving other parts of the brooder in comparative shade, which allows them to choose the light level they prefer.

The number of heat lamps you'll need to heat your brooder varies with the location, type of brooder and the number of chicks being raised; similarly, the height of the heat lamps for a homemade brooder will depend on local conditions

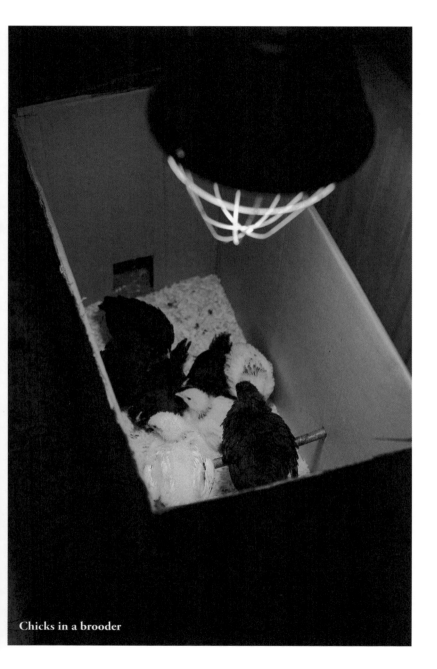

Chicks in a brooder

and the number of chicks. One heat lamp is usually sufficient for a handful of chicks. If you're hoping to raise chicks in a location without electrical power, gas heating is an option. A gas heating element, a reflective hood and a propane cylinder are required, but don't be tempted to assemble your own heating. There are specially designed systems available which are perfect for the job and, more importantly, safe. Choose a well-ventilated yet draught-free space if using gas heating.

Whichever option you choose, make sure you are set up well ahead of time, that the temperature is right and that everything is working well – you want to be confident that your chicks are moving from the protection of their egg into a warm, safe environment. Have spare bulbs to hand!

Avoid heating your brooder from below. Chicks like to get away from the heat source, to regulate their own temperature by moving to cooler areas and underfloor heating prevents them from doing so. You could heat the room the brooder is situated in, but this is expensive as well as energy inefficient, and offers the chicks no cooler area; see Using a Brooder, p.102.

You will also need a small feeder and drinker (see p.106) and allow for some space for the chicks to move around and play.

Incubators

An incubator provides the warmth and humidity that the mother hen would otherwise provide, and you will need one to raise chickens from eggs if you are not using a hen. There are various options for incubators on the market and you should research well before you buy, seeking advice from local trusted chicken keepers and online forums (see Directory, p.246).

A good incubator should have temperature and humidity controls. Capacity is very much a personal choice: ours holds seven eggs. Some incubators have the ability to hold different sizes of egg, which can be useful if you want to graduate on to quail, duck or any of the other home-reared poultry.

As with all equipment, I'd suggest investing in quality – buy the best you can afford and you'll be able to rely on it.

Other bits and bobs

A dust pan and brush of a size to suit your chicken house is essential for cleaning. You may also need a scoop of some kind to deal with the larger deposits created by sizeable flocks. A bucket or wheelbarrow to hold and transport the bedding and waste (ideally to the compost bin) is also handy.

A domestic chicken incubator

Chickens for Eggs

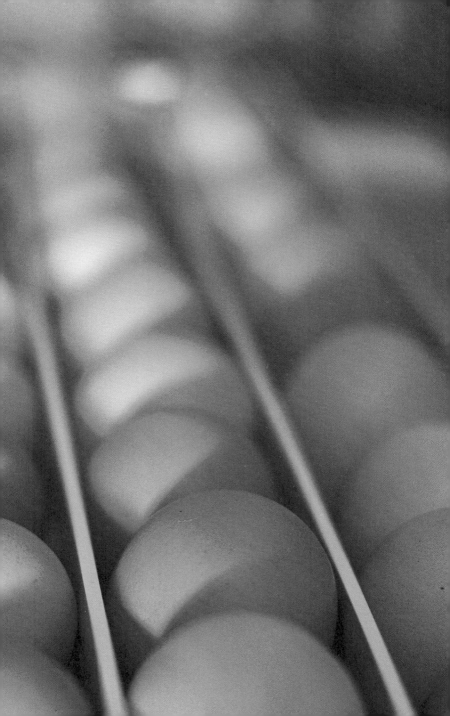

This little everyday ovate object really is something special, in the fullest sense of the word. Within the shell lies everything needed to nurture a couple of cells through an almost impossibly quick process to a live chick. Everything, that is, except the sperm. This is provided by a cockerel, if you have one. If you don't, your hens will continue to lay regardless.

Nutritionally, an egg is deliciously complete. It is an excellent source of protein, containing all the essential amino acids our body requires, with a similar protein level to meat (12 per cent). It also carries a range of vital vitamins, including vitamins A, B1, B2, B5 and B9, and important minerals, such as calcium and iron. About 10 per cent of an egg is fat, but only a quarter of this comprises saturated fatty acids (associated with raising cholesterol levels). Free-range eggs tend to have a higher level of beneficial omega-3 fatty acids and a healthier ratio of unsaturated to saturated fats. The yolk is the most nutritious part of an egg, carrying 75 per cent of its nutritional value. It's hard to argue with the old slogan 'Go to work on an egg'.

You may decide to raise just enough hens to keep you and your family in eggs, or have enough chickens for a surplus from time to time. You're unlikely to have any trouble selling them, giving them away or using them as currency – a dozen eggs has 'bought' us beautiful bunches of flowers, cider, home-cured bacon and a couple of just-caught mackerel. However, you are unlikely to turn producing eggs on a domestic scale into a business. Commercial egg production is highly efficient and sets the price that people expect to pay for eggs. Although some are willing to pay a little more for delicious home-reared eggs, the best you should hope for is to put some money back into the kitty for feed and other supplies.

Choosing a breed for eggs

Layers produce up to 300 eggs a year, declining a little as the years pass. Although the birds are edible, their shape and build doesn't lend itself to a generous dinner.

There is a world of choice when it comes to laying hens and the nutritional value and flavour of the egg doesn't vary from one breed to another. This leaves you entirely free to choose your breed based on their looks, preferred shell colour, laying rate or any other characteristic that takes your fancy. One factor worth considering if you're looking for a lot of contact with your birds is that, for reasons unknown, white-egg layers tend to be more skittish and harder to handle than brown-egg layers.

If you buy a breed that lays brown eggs, bear in mind that the shade of brown can vary considerably even within the same breed, and that as your birds age their eggs often become paler.

Blue and blue/green eggs are laid largely by breeds that originate from South America and these coloured eggs are popular with many. Some of the best varieties are detailed opposite and illustrated overleaf.

BREED	TYPE	EGG COLOUR	LAYING RATE PER YEAR
ARAUCANA A Chilean breed that comes in tailed and rumpless varieties. The birds have distinctive large muffs around the head and a small pea comb. Many colours available – lavender is common. Very placid.	Purebred	Blue	150
COLUMBIAN BLACKTAIL A Rhode Island Red and Light Sussex hybrid that is hardy, lays well and has an easy temperament.	Hybrid	Brown	300
LEGHORN One of the breeds from which battery hybrids were bred. Very productive. Rarely broody. Tame but prefers not to be handled.	Purebred	White	300
BLACK ROCK This is a cross between a Rhode Island Red and a Barred Plymouth Rock. Excellent, healthy free-rangers with a long laying life. Very docile and friendly.	Hybrid	Brown	250
PEKIN A true bantam breed, very tame and small yet robust – excellent in a garden. Pekins lay small eggs in good numbers and have a tendency to go broody, which makes them good mothers.	Purebred	Usually white	180
RHODE ROCK A fine, hardy bird with a long lifespan that lays especially well for the first few years. A cross between a Rhode Island Red and a Barred Plymouth Rock. Very docile.	Hybrid	Brown	300
SPECKLEDY A Rhode Island Red crossed with a Maran, with feathering much like the latter. They are large, elegant, tame birds.	Hybrid	Brown	250
WELSUMMER Beautiful bird with a gold edging to its dark feathers. Layers of large eggs, most are good sitters, calm and easy to handle.	Purebred	Reddish brown	200

Araucana

Columbian Blacktail

White Leghorn

Black Rock

Lavender Blue Pekin

Rhode Rock

Golden Speckledy

Welsummer

Feed and nutrition

All chicks are raised for a month on specialist chick feed, moving on to a pullet grower feed (see p.116) for around 14 weeks until they reach the point of laying. At this stage they move to a layer feed. Ensure you move them from one feed to another, as the make-up of each feed suits that stage of their life. And, rather than an abrupt shift, make each change gradual, mixing the old and new feed over a few days. See p.116 for more detail.

Egg production

The process of egg creation is quietly incredible. When she is born, each hen has all of the ova (egg cells) she will ever produce in her ovary, and at around 18–24 weeks old laying begins.

The ovum leaves the ovary and finds its way to a funnel-shaped organ (the infundibulum) at the top of the oviduct. If sperm is present when the ovum enters, fertilisation can take place. It takes around 3 days from mating for the sperm to travel to the infundibulum, where it can be stored by the hen for up to 3 weeks.

Whether fertilisation takes place or not, an egg is produced and its journey continues the same. It forms from the inside out – the yolk first, then the vitelline membrane and structural fibres, followed by the white (the albumen). The shell, made of calcium carbonate, comes last, forming in the lower oviduct just prior to laying. This whole remarkable process takes just 24 hours.

Contrary to popular myth, brown eggs are not necessarily more delicious than paler ones; the colour of the shell has absolutely no effect on the flavour of the egg. As the hen prepares for laying, she deposits a pigment (that varies with the breed) on the shell. The myth probably originates from the popular pre-war breeds, which lay mostly brown eggs, in contrast to the breeds better suited to post-war mass production, which tended to lay paler eggs.

Structure of an egg

An egg has seven main constituents: the shell, two membranes, the albumen (or egg white), two chalazae, the germinal disc, the yolk and the air sac (as shown in the diagram overleaf).

The shell holds its contents together, protecting them from impact and from harmful bacteria in the outside world, while allowing air to pass through into the egg. Of the two membranes within the shell, the outermost supports the shell by providing some internal structure; the second, immediately inside it, surrounds

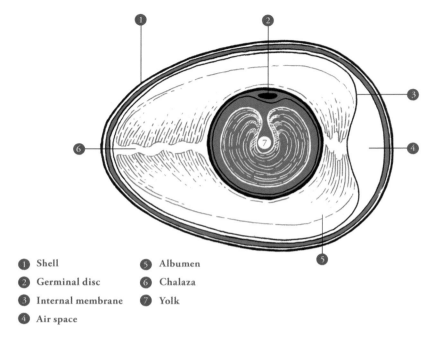

1. Shell
2. Germinal disc
3. Internal membrane
4. Air space
5. Albumen
6. Chalaza
7. Yolk

the albumen. This inner membrane is attached to the yolk by a string at either end known as the chalaza, which holds the yolk and the germinal disc (see below) in the centre of the egg. You can often see it if you look carefully at an egg you've cracked into a cup.

Inside this second membrane lies the albumen (or egg white), which accounts for around 60 per cent of the egg's volume. It is there to provide cushioning and a secondary source of feed for the growing chick. Eggs store well (see p.187), staying edible for 4 weeks or so after laying. Much of this is thanks to the membranes that line the inside of the shell. If these are damaged, infection can occur.

The germinal disc, also known as the egg cell, is where the embryo initiates and grows if fertilisation has taken place.

The yolk of the egg is a vital food reserve for the nearly emerging chick, being absorbed into its body just prior to hatching.

Contrary to logic and, you'd imagine, the preference of the hen, eggs are laid blunt end first. The air sac beneath the shell is at this end and helps cushion the contents of the egg from any bump as the egg is laid. This need to cushion the egg may explain why eggs are shaped as they are, although some believe the elongated oval structure gives the egg strength while causing it to roll in a circle rather than a line, making for fewer losses from the nest.

Laying rates

An egg takes 24 hours or so to make its way from ovary to the outside world. If your bird is particularly efficient at doing this you may get somewhere around 300 eggs a year at their peak; otherwise 200 is more usual, with greater productivity in the longer summer days than in the winter.

Most hens start laying after 18 to 24 weeks, and are most productive until they get to around 2 years old. After this, egg numbers decline and can reduce to none in time. Most domestic owners allow them to peck out their days, adding new young hens to the flock to keep up egg numbers.

It is not only age, breed and time of year that can affect your hens' laying rate; stress can have an adverse effect too. Extremes of heat and cold and unsuccessful predator attacks are among the many stresses that can shock your birds into not laying. Although you should become confident in handling your birds as quickly as possible, don't over-handle them or chase them about when catching them – find a method that works well for you and the bird (see pp.119–22) and stick to it.

Chickens moult during autumn (see p.31). It's a tiring business and egg laying usually decreases in frequency, often stopping altogether; occasionally it doesn't resume until spring. You can stimulate laying with additional lighting if you wish.

Other than that, getting a good number of eggs from your hens is largely about establishing and maintaining a routine and ensuring they have a good balanced diet with access to plenty of water.

Collecting eggs

Most eggs are laid in the few hours after sunrise. Collect your eggs daily, as this ensures you get them at their freshest and that few are damaged. It also minimises the time they are at risk of being eaten by hens or predators. Write on each the date it was laid, using a pencil rather than a pen to avoid the possibility of pigments finding their way into the egg. If you haven't already got one, invest in a container with a soft interior, large enough to hold the maximum haul comfortably. Don't, as I did, wait until the annoyance at spilling yet another egg from your scooped jumper reaches fluorescent levels.

If your hens are entirely free-ranging you may find that they lay in a few outdoor sites, as well as in the nest boxes. They are creatures of habit, so listen for the telltale announcements that laying is taking place. Hens are hormonally drawn to dark, comfortable, familiar sites to lay their eggs. The nest box in their housing provides the ideal place for most, but once in a while a rogue hen will lay at another location. If this happens frequently, it may be that the nest box is overcrowded and you need to provide more laying space. Check that the nest box isn't being dominated by one or two birds up the pecking order – you may have to split the flock if this is the case. It may also be that a bird is sleeping in it and making it too messy to be welcoming. If so, try to discourage it at night and clean out the box more often.

Laying away from the henhouse may be a sign that a hen has a low maternal instinct, or that they just happen to possess the random rascal gene. To resolve the problem, keep their free-ranging to a small area to encourage them to use the nest box rather than their secret spot. Once the habit is broken, the issue seldom recurs.

If you discover eggs and you're not sure how old they are, lower them into a large bowl of water: eggs that sink on their side are likely to be fresh; those that bob upright or float are stale and must be discarded. This is because eggs dry out as they age and the air space inside gets larger, effectively making them lighter and buoyant.

Cleaning eggs

Eggs should always be cleaned, unless you are selling them (see p.72). Even with scrupulously clean living conditions, chicken poo, bits of bedding and even broken egg can find their way onto the shell.

Eggshells are porous but have a coating that reduces the risk of contamination but cross-contamination, such as onto your fingers then onto other food, is a risk. Use slightly warm water to wash them, avoiding soap – it's unnecessary and the scent and chemicals can pass into the egg itself. If you use a cloth, make sure it is non-abrasive and is machine washed afterwards. Dry the eggs with kitchen paper.

Examining eggs

Eggs are generally very similar to each other, varying only in size and colour of shell. However, once in a while, you may find something peculiar.

Rough shells These occur occasionally in most flocks and can be present as rough patches, ridges and so on. They are almost always just harmless oddities and you can eat the eggs, but if they persist, like any oddities, it is worth seeking veterinary advice as to whether it is a sign of an underlying problem with the hen.

Cracked shells You may collect the occasional egg which has a crack in its shell. This can be caused by a drop in the temperature, a hen standing on the egg, or the collector picking it up with a little too much gusto. Cracks can allow bacteria into the egg itself, so you will need to exercise some caution about whether to use it or

not. If it is a very minor crack and/or you are sure that you have just cracked it and will use it very quickly then it is likely to be safe. Crack the egg into a cup immediately before use to check that it looks okay.

Thin shells Soft and/or thin shells often occur when a young bird begins or nears the end of her laying life, and occasionally in extended periods of hot weather when the hen's metabolism increases. At other times it usually indicates a mineral deficiency, which can be alleviated by providing soluble grit (see p.116), but it can also be a sign of disease (see p.152) so be aware if the problem persists.

Odd shapes and colours Unusual colours and shapes are part of the natural variations that come with collecting eggs every day. They are almost always entirely fine and perfectly good to eat, but odd shapes are not the best for a developing embryo so don't use misshapen eggs to hatch into chicks.

Fertile eggs It may seem obvious, but without a cockerel there can be no fertile eggs. With a cockerel in your flock of layers there is the potential for fertile eggs, but unless the embryo has been growing for a few days there is no way of telling. Fertile eggs look the same and taste the same as unfertile eggs. If you haven't been collecting the eggs daily and/or you want to keep back any fertile eggs to hatch, you can check the eggs by candling (see p.95). Be aware that even if the egg was fertile, if you put it in the fridge it won't be any longer.

Blood spots Little dark spots in the yolk, known as blood spots, are fairly common. They are entirely harmless and not (as some believe) a tiny embryo beginning to form. They are generally caused by one of the minuscule blood vessels in the ovary breaking as the yolk is released. Blood spots tend to be more common if your hens have been disturbed in some way (if moved to a new home, for example).

Meat spots These dark spots are found in the white rather than the yolk and are often darker than blood spots. They are usually tiny pieces of the wall of the oviduct and are harmless.

Double yolkers Once in a while you may find an egg with two yolks. Statistically this happens once in around a thousand eggs, when two ova pass through the ovary simultaneously, but certain breeds (especially young, highly productive breeds) and some individuals seem particularly prone to it. It can even happen more often at certain times of the year. One of our random hybrid crosses lays double yolkers relatively often in spring. There is nothing wrong with them at all – think of them as a bit of a bonus, especially if you're planning to make an omelette.

Yolkless eggs At the beginning and occasionally at the end of a hen's laying life, you may find the odd egg without a yolk. Also known as 'wind eggs', these are usually just a passing phase and nothing to be concerned about.

Soft shell This is almost always caused by a lack of calcium in the hen's diet. Crushed oyster shells (see p.116) are an excellent source of this essential mineral and some brands of layers' pellets have good levels. You should also ensure your hens have some fresh green vegetables in their diet, as eating a good proportion helps the chicken to absorb calcium more readily. A shock incident, such as moving home or a predator attack, can also result in soft-shelled eggs for a short while. Worms and old age are other causative factors.

Unusual yolk colour A hen's diet can have a considerable effect on the colour of the yolk: grass and clover deepen the yellow, whereas a lack of these can make the yolks paler. In early spring you may notice that the odd yolk is green: this suggests too much lush new grass in their diet and you should increase the proportion of grain/pellets you're feeding them. Very occasionally, an unusual yolk colour flags up a health issue. If your hen suddenly starts laying markedly lighter-coloured yolks for no obvious reason, keep a watch for any signs of illness. If there is grey, brown or mottling to the yolk, it is likely that the egg isn't fresh, so discard it.

Rotten eggs Be a little wary of any eggs you find outside the normal areas you collect from. Occasionally a hen will start laying somewhere different and you may come across an egg or clutch of eggs whose age you can't be sure of. Chances are they will be perfectly fine but keep them separate from any others you've collected and crack them into a cup just before using, to be on the safe side. Cracking open a rotten egg is all you'll need to do to confirm that it is past its best – it will stink to high heaven.

Watery white Although you have the best part of a month to eat an egg if it is stored well, the quality begins to decline from the moment it leaves the chicken, and a watery white almost always indicates an old egg. Occasionally it occurs in eggs laid by a bird reaching the latter stages of its laying life and even more rarely it may indicate a virus in the bird. Discard the egg and keep an eye on your flock for any signs of illness, though it is unlikely that there is anything to worry about.

Other problems Very rarely you may find something else in the egg: the odd scrap from the reproductive canal, or a parasitic worm may have found its way in. Neither has happened to me in a decade of keeping chickens, but it is possible. If you're in any doubt, discard the egg, and if it is a worm, seek veterinary advice.

A broody Orpington with her eggs

Broodiness

In the hour or so before an egg is laid, a brief surge of hormones causes the hen to seek out the nest, to peck the bedding about a little and to sit. After laying, she is likely to get up and on with her day. Occasionally, in summer when the days are longest, a hen's hormone levels remain high and the normally passing urge to nest remains. This is what is referred to as a 'broody' chicken.

When a hen becomes broody, an instinct drives her to want to hatch and raise young. Mating need not have occurred and the eggs need not be fertile; indeed even when there are no eggs or cockerels broodiness can occur. The hen will sit in the chicken house for most of the day, rising only briefly to eat, poo and drink. Her breast will be unusually hot and she may make disgruntled noises and fluff up her feathers if you try to move her or remove any eggs – she may even peck at you. Take her out of the house and she may well just sit low and complain.

If you're looking to breed chickens using a hen (see p.136) then the tendency to broodiness is a godsend, otherwise it can be a nuisance as your hen will stop laying and may become isolated from the flock during the time she is broody.

Broody behaviour generally lasts around 3 weeks or so, but it can go on longer. If it becomes a nuisance, you can try to deal with it. Despite the hen's typically noisy protestations, remove the eggs every day. This often curbs her behaviour after a few days. One of our henhouses has a roof that lifts up, and if a bird becomes broody in this house we open the roof as this helps to cool down the heat that comes with broodiness.

Sometimes a broody hen will continue to sit, even if no eggs are present; this is caused by the drive to incubate. If you let her sit out the 3-week period after which the eggs would have hatched (checking daily to make sure none are present), then she will almost always give up.

Once in a while she will either continue to sit or restart the broody behaviour soon after ending it. You can nudge her off her sitting spot, let her carry on her harmless (if unproductive) behaviour, be ruthless and replace her or slide a few fertile eggs under her and take advantage of the situation.

Egg-laying breeds (see p.59) rarely present a problem with broodiness. It is more likely to occur with dual-purpose breeds and one or two are particularly susceptible, such as Buff Orpingtons. It is a good idea to bear this in mind when making your choice of breeds.

Occasionally broody behaviour can become more of a problem. For example, if the hen is sitting on a secret stash of eggs away from the nest box she may be reluctant to leave them at night. It is advisable to lift the hen off the eggs and relocate her and the eggs to the nest box if you want her to incubate them, or just remove the eggs.

Legalities of selling eggs

Pleasingly and perhaps unexpectedly, there are few legalities and guidelines when it comes to selling your home-raised eggs from your doorstep. It is, however, important to follow them to the letter.

- You may not sort your eggs into different sizes or grades of quality. In practice this often works in your favour as many people like having a range of sizes to choose from for different requirements.

- The eggs should be clean but not washed as this can remove the protective bloom on the outside of the shell that helps prevent bacterial contamination. The shells should be without cracks.

- It is your responsibility to ensure that the eggs you are selling are fresh for sale and consumption, so sell them within a few days of laying and keep a strict system whereby your eggs are stored in date order of laying so that they are sold in order too. Writing the date of laying on each egg with a pencil helps greatly.

- You may sell eggs to individuals but not to an organisation or establishment, such as a restaurant or caterer, that will pass them on to someone else, either raw or cooked.

- You may sell at a market direct to consumers, but you must also provide your name and address, a 'best before' date not more than 28 days after the eggs were laid, and advise customers to store their eggs chilled after purchase.

- Unless you have had an inspection by the relevant DEFRA department, and if necessary a certifying body such as the Soil Association, be wary of using terms such as 'organic' or 'free-range': they are legally defined terms that impose certain requirements, including inspections and paperwork.

- If you're tempted to produce eggs on a commercial scale, be aware that there is considerable red tape to comply with. To familiarise yourself with the legalities, check the DEFRA website (see Directory, p.246).

- If you have 50 chickens or more at any point in the year, regardless of whether you intend to sell the eggs or meat, you must register your flock with DEFRA.

FREE
TO
ROAM
EGGS ON SALE

$1

PER DOZEN

Chickens for Meat

Homegrown chicken is extraordinary.

Typically raised more slowly than a commercially bred chicken, it has the chance to live longer and often more happily, and you will be able to taste this difference when you come to eat it

There are two main approaches to raising your own chicken for the table. You can choose fast-growing commercial hybrid breeds, feed them food designed for mass production and finish them quickly – this will give you good chicken in a relatively short time. Alternatively, you can raise chickens from breeds more suited to non-intensive rearing, feeding them on a mix of organic chicken feed and natural forage – giving your birds a longer life and usually giving you a leaner, less fatty carcass than is available in the shops.

Both methods are valid and interchangeable, and you can do both at the same time if you wish. I have to say I only follow the second, more extensive approach partly because I prefer the flavour and texture of the meat, but equally because I enjoy raising them in a less business-minded way and giving them a longer life.

Choosing a breed for meat

Most meat birds have been bred specifically to take up more room on your table A larger frame, bigger breasts and a speedier rate of growth make them ideal if you're after meat rather than eggs. Meat breeds tend to be poor layers, though you can expect a few bonus eggs.

Providing it is fit and healthy at the time of its death, you can eat a bird of either sex of any breed, but some suit being raised for meat more than others.

Most of the chicken available in the shops comes from Cornish White Rock hybrids that have been selected to perform well and produce good meat quickly Intensively reared birds can hit the table in little over a month, with the fastest outdoor-reared varieties ready to eat in around 12 weeks.

Intensively reared varieties aren't really an option for domestic keeping as they have been bred to be fairly inactive and are prone to ailments when free-ranging Although free-ranging birds of these varieties are available, I recommend slower growers that take 6–9 months to mature. Of the breeds in commercial outdoor production, we've reared Hubbards at River Cottage, allowing them a longer, less intensive life than they'd usually have and the meat was outstanding.

As their name suggests, dual-purpose breeds lay fairly well while growing to a good size for eating. This makes them popular with many small-scale chicken keepers, as any excess chickens make good eating. The eggs taste the same as layers eggs and the meat is as good as you'll get from meat birds, but the price for this appealing duality is fewer eggs than the average layer and generally slower-growing smaller birds than the true meat breeds.

There are many meat and dual-purpose breeds to choose from. I recommend that you consider the varieties detailed opposite and illustrated overleaf.

BREED	CATEGORY	GROWTH RATE	LAYING
ORPINGTON Calm, heavily feathered birds. Good sitters and mothers. Buff Orpingtons are the most popular, but there are black, blue, white and golden varieties too.	Dual-purpose	Medium	Good
HUBBARD A fast-growing commercial breed, often ready to dispatch in around 12 weeks. Excellent free-rangers, usually calm, not easy to breed true from.	Meat	Fast	Poor
WYANDOTTE Large, well-feathered birds, with a placid nature. Tendency to broodiness. Very good mothers. Various colour patterns, including Barred, Gold Laced and Buff.	Dual-purpose	Medium	Good
BARNEVELDER An old, calm, easy-to-keep breed that lays dark brown eggs at a reasonable rate while growing to a good eating size. Comes in many colours, including black and white.	Dual-purpose	Slow	Good
RHODE ISLAND RED A large bird with a heavy feather covering even down to the feet – a good choice for cold areas. Easy to look after and handle. Many colour options. Excellent sitters.	Dual-purpose	Medium	Very good
PLYMOUTH ROCK Buff, blue and barred (black-and-white-striped) varieties of this old American breed are the ones to choose for both eggs and meat. Calm birds, good sitters and excellent free-rangers.	Dual-purpose	Medium	Good
MARAN Diverse breed: varying colours, degrees of broodiness and calmness. Layers of dark brown eggs. Slow-growing, but their meat has an excellent flavour.	Dual-purpose	Medium	Good
SUSSEX Reliable sitters, calm and easily handled; the males in particular make good eaters.	Dual-purpose	Medium	Very good

Orpington

Hubbard

Partridge Wyandotte

Barnevelder

Rhode Island Red

Buff Plymouth Rock

Maran

Light Sussex

Keeping meat birds

The process of keeping chickens for eggs or for meat is very similar but there are some important differences. After leaving the mother or the brooder, chicks need to be raised separately from their parents and away from laying birds, as their food requirements are different (see p.116). It is crucial that they have feed specifically for meat birds – it will be high in protein and allow them to grow at a rate that enables them to mature to a good size for the table. Meat birds also need constant access to feed.

Grass can form a good part of a meat bird's diet and is especially good if you can strip graze your birds, i.e. rather than allow them unrestricted access to a large space, keep them to a smaller area, moving them on to fresh grass regularly before the grass is eaten too thoroughly or grow-back can be poor.

You can use the same type of housing as for layers, although perches and nest boxes aren't necessary for meat birds.

Choosing the sex

There is no 'better' sex to raise for eating. Most hybrid chicks are supplied unsexed as they are equally good for eating and mature at similar rates. Away from these hybrids, many domestic keepers raising dual-purpose birds from chicks or eggs raise the cockerels to eat, rather than dispatching or selling them.

How many to raise

I'd recommend a small number for your first batch. Then, if you feel the need to up the size of your flock, you can do so confidently, having gained experience from handling an easily manageable number. A dozen is a good start-up number that allows you to become accustomed to the process while having a good-sized 'harvest' at the end. It also enables you to see how this scale works for you and, equally importantly, your storage capacity. A dozen frozen chickens takes up a fair bit of freezer space.

If you want to raise all the chicken meat that you and your family eat, working out how many you'll need is a straightforward case of multiplying the number you get through in a week by 52 weeks of the year. However, don't let this pressurise you into raising all the chicken you consume. It is crucial that you don't exceed the number that suits you and your set-up.

Generally speaking, I'd recommend raising the total number of birds you want in more than one batch. For example, if you want 52 chickens, so that you can eat one a week, raise them in four batches of 13 birds. If disease or predators hit, you can lose most or all of your birds very easily, but splitting the numbers into smaller batches keeps the risk of a total clear-out to a minimum.

Several smaller-sized batches are usually easier to deal with too. The larger the batch the bigger you'll need your brooder, housing, dispatching, dressing and storage/freezer facilities to be. It is also worth bearing in mind that although there are some economies of scale when raising larger batches of birds, your outlay comes in bigger lumps.

Feed and nutrition

As for layers, meat birds spend their first month feeding on chick feed, moving on to a higher-protein feed (see p.116). If you are providing them with good-quality feed and they have access to grass then they are unlikely to require any dietary supplements. Remember, their lives are typically much shorter than the lives of hens kept for eggs, so vitamin deficiencies are unlikely to have time to develop. Do keep your meat birds separate from your laying hens – meat birds need access to a different feed that is unsuitable for layers. For further information about feeding chickens, see pp.113–19.

When to get meat birds

You can raise chickens to eat at any time of the year but there are some practical considerations that may influence when you do it.

Some breeds are hard to source in autumn and winter, and if breeding using parents from your flock, be aware that they are usually less fertile during these seasons. That said, if you order chicks well ahead, you should find most varieties are available.

Other considerations relate to your local conditions. Ideally chickens should move from the brooder to outdoor life when it is relatively dry and neither very hot nor cold. This isn't always predictable, but avoiding a wet site and the height of the summer (or providing adequate shade) makes good sense.

April is a particularly good month for chicks to reach independence, as the days are lengthening and warming and there is grass aplenty. Also, be aware that meat birds will grow more slowly and eat more food in cold periods, using up more of their intake in keeping warm.

All things being equal, if your birds are ready to move outside in mid-spring/early summer, with a late-summer or autumn dispatch (depending on breed), this will give your birds the best access to good grass, the potential for dry conditions, and cooler temperatures when being slaughtered, plucked and dressed. This means thinking ahead and allowing time, depending on whether you are starting with eggs or chicks. You can also start with older meat birds, but some or most of the 'growing' will have been done for you and they are likely to be costly.

A group of Light Sussex hens

Judging when meat birds are ready

As with vegetables and fruit, meat birds have a point at which they are perfect for 'harvesting'. With most of them this is at around 18–20 weeks. Hybrid varieties (see p.23) often reach this stage sooner and are vulnerable to sudden death if they are allowed to over-mature. In general, you can expect more strongly flavoured meat the longer you leave meat birds, and some commercial breeds may become less tender after their peak. Of course, if you prefer that texture/flavour combination, by all means give them a little longer.

If you are raising birds to eat, you need to be able to judge the right time to dispatch them. You can eat a chicken at any age, so it's more about picking the optimum time for you. If you are looking at it with a commercial eye, you'll want to slaughter them when they have grown to full size and little beyond. For some meat birds this can be as young as 10 weeks after birth, but for most domestic keepers it will be considerably longer.

It is very much a matter of judgement. If you've handled your birds throughout their life you will be aware of their changing shape and size and be able to judge when it is the right time. As a rule, when a bird starts laying or crowing it's reached a prime size. You can keep your birds longer, but the meat is likely to become less tender, if more flavoursome. It really is about personal taste. Try your first birds when they reach this point. You can always leave the next ones longer if preference, inquisitiveness or circumstances dictate.

If you have a few cockerels maturing at the same time and you're experiencing (or anticipating) the start of conflict, you might consider dispatching a few a little early. This also gives you a steadier supply than getting them all at once.

Never let commercial meat hybrids go beyond 18 weeks. By this time they should be large and ready; letting them grow further risks them reaching a size that their legs and heart cannot support.

Starting a Flock

Getting your first chickens is a wonderful moment.

It's a bold step from the world of growing and gardening into the world of livestock – living creatures with personalities and peculiarities. They move! Exciting as it is, it is also quite a responsibility to look after living creatures, so take a moment to consider how you want to begin your chicken keeping. You can start a flock with eggs, day-old chicks or older birds. Each of these options has pros and cons (see right). Most of us start with adult chickens, perhaps moving on to raising chicks or hatching eggs with the benefit of confidence and experience. It's easy to source adults from breeders, fellow domestic chicken keepers and other suppliers, and if you're in any way anxious about raising young birds or hatching eggs, starting with adults is the way for you.

Adults are available at a range of ages, but many people opt for buying 'point of lay' birds, aged around 14–18 weeks, as they are likely to be about to embark on the most productive laying year of their lives. Another option is to buy pullets. These birds are usually older than a point of lay bird but less than a year old – the advantage of buying at this age is that you can be relatively certain that they are already laying well, yet not past their laying peak.

Welsummers at auction

	PROS	CONS
STARTING WITH ADULTS	Potential of immediate eggs Easy to examine before purchase You can be sure of the colour/size Easy to identify the sex Less fragile than eggs and chicks No additional equipment	Potential uncertainty of the age of the birds Higher costs of birds, compared with eggs/chicks Potential of disease in older birds, due to their typically long exposure to many birds
STARTING WITH EGGS	Potentially cost-effective if raising large numbers The pleasure of raising birds from eggs Allows you access to more breeds Enables you to add to the flock by hatching in the winter when few birds are laying	Longer wait until you have eggs to eat than if buying adults Uncertain sex – 50 per cent are likely to be male The issue of what to do with unwanted males Uncertain fertility – some eggs may not be fertile Requires incubator and brooder
STARTING WITH CHICKS	Potentially cost-effective if raising large numbers The pleasure of raising birds from very young Fewer losses than when raising from eggs You start with the number of adults you want, compared to starting with eggs	Longer wait until you have eggs to eat than if buying adults Requires a brooder Uncertain sex with most breeds at one day-old – 50 per cent are likely to be male The issue of what to do with unwanted males

Buff Plymouth Rock hens

Starting with adults

Buying adults is fairly straightforward, and if you do your homework first you are likely to make a great start in chicken keeping. Get to know a good, well-respected local breeder and the chances of anything negative will be minimised. Ex-battery hens (see p.20) are usually moved out of commercial production when first moulting and can be back to looking and being healthy within a couple of months. You can source these from the British Hen Welfare Trust (see Directory, p.246).

Some tips for choosing chickens:

• If possible, avoid buying birds in autumn, which is the moulting season. Apart from not looking at their best, the added stress of a change of home can weigh heavily on them, leaving them susceptible to illness.

• Ask for the breed, sex, age and colour of the birds you are buying – preferably in email or other written form, especially if you're not buying them in person.

• If possible, hold the bird(s) and examine them properly before buying: you may feel nervous of doing this at first but you'll need to get used to it. With experience you'll get a sense of whether they are too large or scrawny.

• Check for wounds and/or large bare patches (small patches from mating are quite normal).

• Check the face. The birds should be breathing easily, with beaks closed and with no rattling. There should be no nasal discharge or runniness to the eyes.

• Look through the feathers for parasites (see p.150). The presence of lice or a white deposit should persuade you to keep your wallet in your pocket.

• Check their rear end – the vent, under the tail. It should be fairly clean.

• Toes should be straight, all present and the comb should be undamaged.

• Most adults are easily sexed by colour, but some breeds (usually white or black ones) need a closer inspection. Comparing male with female will show that the cockerel has pointed rather than rounded feathers on its hackle, its comb and wattle are larger, and it may have iridescence in its neck and tail feathers. Obviously, the ability to lay eggs or crow is a bit of a giveaway too.

- It is difficult to age an adult chicken. Hens lay well until 4 or 5 years and cockerels are fertile over a similar period but both can live for around twice that, so the risk of buying unproductive chickens is clear. An egg-laying hen usually has a dark red, shiny comb and wattle, whereas ageing hens or cockerels will have less bright, matt combs and thicker scalier skin on their legs. There are other signs of age in hens – narrow pubic bones and a moist cloaca among them – but these are not for the novice.

Making a new flock feel at home

Exciting as it is, don't be tempted to gather family and friends around your new birds as they scramble out of their carrier and into their new environment. They'll be anxious, possibly stressed by the journey and they'll need time to adjust. The best transition they can have is to be gently placed in their new house with some food and water and locked in overnight. Without the sensation of travel they will calm down, rest overnight and emerge the next day having become at least partially accustomed to their new home.

On the first morning, open the door of their house and let them emerge in their own time. Usually they will pop straight out, but nervous or ex-battery hens may take longer. A little food and water outside the door can encourage them.

Sit with your new birds when you feed them and they'll get accustomed to you, but it is best to keep others, and in particular any pets, at a distance early on.

Introducing new birds to an existing flock

Losing birds from and adding birds to your flock is inevitable, but it brings with it a certain level of activity within the flock. The hierarchy, or pecking order, becomes unsettled and takes time to re-establish. If handled badly, adding new birds can result in violence and even death within the flock, so it pays to be prepared:

- Ideally keep new birds in quarantine for a fortnight, to examine them for lice and to allow any other ailments to become apparent.

- Never introduce a cockerel to a flock that already has a cockerel – they are likely to fight and one or both may die.

- If you plan to add hens to the flock, add them in groups so that any potential bullying that may occur as the pecking order establishes is diluted.

- If you can, put the new, quarantined birds in an enclosure next to the main flock so that they can become used to each other's presence and the hierarchy becomes fairly established before they are mixed.

- Provide something for birds to hide behind or under. Not only does it give any persecuted birds a break, the act of hiding displays submissive behaviour that helps establish the hierarchy.

- If possible, move the entire enclosure at the same time as you introduce new birds – it often results in a more gentle acceptance of the newcomers.

- Add new birds to the flock at night, in the house, and be on hand in the morning to check that any of the 'getting to know each other' isn't violent.

- Some fighting is likely, but only intervene if an altercation is serious. Remove any bleeding birds as this can lead to a frenzy of attacking and potential death.

- Even when the disruption seems to have settled down, keep an eye on your birds to ensure they are all getting food and water. Putting in extra feeders and water containers will help ensure all birds have access to what they need.

- If all else fails and there is no calm after a week or so (thankfully very rare), you will have to consider housing some or all of the new birds separately.

Quarantined birds

Starting with eggs

You may want to begin your adventures in chicken keeping (or add to an existing flock) by raising them from eggs. It is a reasonably straightforward undertaking but it requires some equipment and adherence to a fairly precise timetable.

You will need an incubator to provide the warmth and humidity required to hatch the eggs, and a brooder in which to raise the chicks for their first few weeks.

Sourcing fertile eggs

If you have friends that keep chickens, they may be willing to provide eggs that, thanks to the presence of a cockerel, may well be fertilised. This is a fine route to take as there will be minimal storage and transport time involved, reducing the potential for breakages, or eggs being kept at the wrong temperature for any length of time. Local suppliers may also be able to help.

Otherwise you can source fertile eggs from a commercial supplier. This is a good option if you are looking for particular breeds, especially heritage or rare breeds. Wherever you source your eggs from, 50 per cent are likely to hatch out male and there is rarely a guarantee that any will be fertile or hatch, so do be aware that there are few certainties when starting with eggs.

Using an incubator

If you are hatching eggs to start your flock, you will need an incubator. If you're hatching eggs to add to an existing flock, you can use either an incubator or a broody hen (see p.136). You will also need a brooder to raise the chicks (see p.102) if you're not using a hen. Most incubators are easy to use, but they vary, so follow the instructions. Keep your incubator spotlessly clean and sterilised at all times. Incubators have some advantages over hatching using a hen:

- Some broody hens can occasionally and suddenly lose interest halfway through sitting.

- Hand-raised chicks tend to be friendlier and more attached to you than those born under a hen.

- You can use an incubator at any time of year, whereas many hens will sit only in the warmest months.

- An incubator can hold more eggs than a single broody hen can hatch.

- It is easier to monitor the eggs more often than when hatching under a hen.

Adding eggs to the incubator Turn the incubator on a few hours ahead of adding the eggs to allow it to get up to the correct temperature and humidity. This also enables you to identify any problems before the eggs are added. When the incubator is ready, add the eggs, taking the following steps:

• Place the eggs in the incubator on their sides or with the blunt end up, depending on the model you have (check the instructions).

• Take great care when handling the eggs. Always wash your hands thoroughly beforehand and handle the eggs gently to avoid damaging the shell or the embryo within.

• If you are hatching more than one breed, keep each breed together. If the breed isn't immediately apparent from the shape, size or colour of the eggs, draw a diagram showing the position of the different breeds.

Temperature The temperature level is critical to successful hatching. You should keep the eggs at 37.5°C for the initial 18 days, then reduce it just slightly to 37°C until the chicks hatch.

Humidity This is also pretty crucial. A small air bubble sits at the base of the egg, cushioning the embryo and the other contents of the egg from stresses, especially temperature fluctuations. If the humidity is too low and the air is dry, the moisture in the egg evaporates quickly. Consequently the air bubble expands and the chick has neither the room nor the fluid it requires. Conversely, if the humidity is too high, then the air bubble, which acts as a reservoir for the chick to breathe, can be too small – the chick can literally drown in trying to be hatched.

A broody mum seems to get the humidity required at each stage of development just right. To mimic this in your incubator, set the humidity to 53 per cent (give or take a degree) for the first 18 days, upping it to 70–75 per cent until hatching.

Turning The mother turns the eggs regularly, especially early on. This prevents the developing embryo from sticking to the membrane that lines the shell and reduces the risk of problems such as deformity. Some incubators turn the eggs for you. Otherwise, you should give them a quarter-turn around five times a day, stopping on the eighteenth day. Remember to turn the eggs in different directions. Turning them in one direction every time can cause one chalaza (the string holding the yolk in the middle of the egg) to twist tight and the other to unravel. Place the eggs on their side, with the blunt end slightly higher than the pointed end. Marking the eggs with a pencil helps to remind you which have been rotated.

Candling

Candling is the simple process of shining a light on an egg to ascertain whether the egg is fertile and to see how the embryo is developing. It has to be carried out with consideration for the embryo, i.e. with speed, care and without changing the temperature of the egg. It is, after all, fairly invasive and startling – like shining your headlights into someone's bedroom when they're sleeping.

It is best to carry out candling in a warm, darkened room. You should do it quickly and only once or twice for each egg: look at the developing embryos after a week, re-checking only the ones you are uncertain of 3 or 4 days later. You should be able to identify the air sac and, if the egg is fertile, a network of veins. Immediately remove any eggs that are not developing, along with any that seem to be swelling or sweating. You can buy a specialist candler or hold the egg a few inches from a 60W light bulb. The light shining through the egg casts shadows where you should be able to identify the embryo, air sac and even the blood vessels.

Candling can be carried out on eggs that are under a hen – lift each egg out swiftly, one at a time, check quickly and replace quickly to minimise upsetting the sitting hen – but obviously it is much easier to do when hatching with an incubator.

A fertile egg

1. Network of veins
2. Embryo
3. Air sac

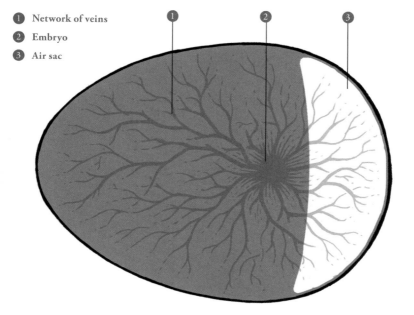

Hatching and the chick's first days

Whether the eggs are under a hen or in an incubator, the hatching process is the same. With the correct temperature, humidity and turning, it will be just 18 days until the chick begins to peck its way into the air sac using a small 'tooth' at the end of its beak. The hen, if there is one, may well encourage the chick with gentle clucking. Over the following 3 days the chick will make a small hole in the shell. At this point it may rest, to allow the remaining yolk to be absorbed into its body to sustain it for the next couple of days. That completed, the chick will peck a hole in the shell or a line of broken shell large enough to push through.

The chick will be wet and possibly shaky as it emerges from its shell. Leave it in the incubator to adjust, dry and become fluffy. Having absorbed the yolk, the chick will not need food for a couple of days. If you are using a hen, it will take care of the chick; if you are using an incubator, move the chick to the brooder (see p.102).

Clean your incubator thoroughly as soon as all the chicks have hatched.

Incubating problems

Often, not all the eggs will hatch. This is to be expected: some eggs may have been infertile, or not developed past the early stages, or been jiggled too much or kept at the wrong temperature too long in transit. If repeated problems occur, or a batch shows a particular trend, it may be attributable to something that can be rectified.

If the temperature of your incubator is too high, it can lead to premature birth; if too low, birth may be delayed. Both eventualities can result in unhealthy chicks.

If hatching occurs more than 24 hours after 21 days in the incubator, it is likely that your eggs were incorrectly stored prior to incubation or that the incubator isn't consistently warm across its surface.

If eggs frequently don't hatch by day 22, check inside a few: lack of an embryo suggests that the egg was infertile; a small embryo indicates incorrect temperature or humidity in the incubator, or that the eggs weren't turned frequently enough or possibly too often, or that the developing chick succumbed to a bacterial infection.

Malformed chicks can result from too low a temperature, humidity being too high, disease or insufficient turning.

Chicks with bloody, crusty and/or red, weeping navel areas may have omphalitis. This is caused by bacteria, and could be a reminder to improve the hygiene of your equipment, and the surfaces on which it is housed, and to wash your hands before and after handling. It may also be an indication that your incubator temperature is too low and the humidity too high.

Occasionally, you may feel that a bird is taking too long to break through or struggling to leave the shell – it is almost always inadvisable to help out, as the chick rarely survives. If you feel you must intervene, do so as little as possible, helping to remove a little shell with sterilised tweezers.

An emerging chick

A few-day-old chick

Starting with chicks

You can also start with day-old chicks. The upside of starting with chicks is that you begin with the number of birds you want (rather than the number of eggs, which may not all hatch), you'll be a few weeks nearer to having eggs to eat and it saves you the expense of an incubator.

The single largest pitfall of starting with chicks is that it's easy to forget that they grow into adults. They are cute and lovable, they don't take up much room and adding another few to your flock seems entirely manageable. Then they grow, along with their food bill. It is advisable to stick to your requirements. Also, do remember that unless you are buying chicks of one of the breeds that can be sexed immediately and therefore guaranteed by the supplier, the law of averages means around 50 per cent are likely to be male.

Chicks can be sourced either in person or by mail order from specialist hatcheries (see Directory, p.246).

Buying chicks in person

If you can, buy your chicks in person. Sourcing from a local breeder means you get to see the chicks in the environment in which they've been raised and you may also be able to buy in small numbers. If you have a good relationship with a chicken or equipment supplier who also supplies chicks, consider them too. Alternatively, if you happen to know someone who is hatching birds, they may be happy to add a few eggs to their batch for you. Nothing rivals a trusted recommendation for a good local supplier, but in the absence of one, check forums and other websites for recommended suppliers.

It may even be possible to buy chickens that are a little older – between the newly born and adult stages – so that you can ease your way gently into raising young chickens.

One potential limitation of sourcing from a small-scale supplier is that they often have a limited range of breeds and the young chicks may only be available for part of the year. Also, vaccinations (see pp.124–5) are not carried out by all small-scale breeders, so do check.

Some tips for buying chicks:

- Think about the timing. You can buy chicks at any time of year, but I'd advise you to avoid buying just before winter. Chicks take around 4–5 months to begin laying, but most breeds will lay regularly only when the days are relatively long, so avoid them maturing in winter. Similarly, for meat birds, when they emerge out of the brooder after a month or so, you're likely to want them to roam freely on some grass, which can be in short supply in winter. If you want to get chicks in winter, then February is a great time to source your chicks – the chicks emerge from the brooder into longer days, the first grass coming through and a long season ahead.

- Source and order rare breeds well ahead of time, as there is often great demand for a limited number of birds.

- Chicks vary enormously in colour, depending on breed and sex. Familiarise yourself with the colour you expect your chicks to be. Your supplier's website may carry details, otherwise search a few of the websites in the Directory (see p.246) and/or ask questions on the forums. Getting the colour right helps avoid buying the wrong breed.

- Determining male from female for most newly hatched chicks is practically impossible. The only exception to this is if they are sex-linked chicks. These hybrids, such as Cream and Gold Legbars, have male offspring of a different colour to the female young and you can distinguish male from female on day one. A few breeds have different-coloured wing tips in the male and female, which can be used to identify them at around 10 days old. After a few days an experienced breeder can examine the cloaca to determine the sex of the chick, but it is time-consuming and delicate work, hence the premium paid for sexed chicks when buying them from a supplier.

- As the chicks grow, distinguishing between the sexes becomes easier, because differences in characteristics such as tail feathers and comb growth become more apparent. Often, even slight differences in attitude, behaviour or growth rate can give you an inkling of which may be cockerels, to be confirmed (or not) as they develop. Wattles and combs often develop more quickly in cockerels, becoming more richly coloured and defined at 3 weeks than in females. After 2 months or so, size, wattle and comb, and tail length will usually tell you if a chick is a cockerel.

- Although small, you should be able to see the comb on most chicks and to distinguish whether the mini-comb equates in form to the shape it should have for that breed.

- Check whether the breeder has its birds tested and vaccinated against major ailments, as many of them do. You may be offered the option of vaccinations, especially for Marek's disease (see p.152); it is worth the extra expense.

Chicks in a homemade brooder

Buying chicks from a hatchery

Specialist hatcheries prepare eggs and hatch them out to order. The day-old chicks will usually be sent to you by post, though there is normally a minimum order of a couple of dozen or so. Chicks can survive the day's postal transit as the yolk that clings to them soaks into the body and sustains them for a couple of days, and the combined warmth of the group keeps them at a good temperature. Always open any deliveries immediately and check the chicks at once. If there are any fatalities or if anything is less than perfect, inform your supplier immediately.

As orders are received, breeders and hatcheries add more eggs to the incubation process and in around 3 weeks your chicks are born. Bear this time lag in mind when you are planning.

Caring for newly arrived chicks

Whether you're buying chicks in person or from a hatchery, their arrival should be predictable. Have everything ready before they appear at your door.

When the chicks arrive, allow them some time in their brooder, undisturbed, to relax with food and water (see p.104), and they should calm down from the anxiety of their journey quickly. Check the eyes for clarity and brightness, and the backsides for cleanliness. Toes and beaks should be present and straight.

Heat is a key factor in a chick's behaviour. Baby chicks are either noisily carrying on with life, or asleep. Panting is a sign that they are heat stressed, so cool them down by giving them more room and/or moving any source of heat a little further away; clustering together can mean they are too cold.

If chicks don't stop panting when cooled, or they are excessively noisy, they could be ill. You may see a small lump on a newly hatched chick's belly – this is where the egg yolk was and it is perfectly normal. Other redness or wounds may not be a good sign. Look for chicks that are acting differently from the rest of the group – perhaps slower or listless – they could be unwell.

If you have any concerns and/or are unable to satisfy yourself that the chicks are healthy, don't buy them or if they have been delivered, register your concern with the supplier.

For handling chicks, see p.122.

Using a brooder

If you have hatched chicks using an incubator, or have sourced day-old chicks, you will need to make sure they have the warmth, security and shelter that the mother hen would naturally provide and this is best supplied by a brooder. The chicks also require suitable food and water to sustain them as they grow.

The perfect temperature for your chicks in their first week is 35°C. This should be lowered by 5°C each week until the temperature inside and outside the brooder is the same. The temperature should be measured about 5cm above the level of the bedding and the heat source raised or lowered until the required temperature is reached. An easy-to-read digital thermometer is often best for this.

If you are using an infrared heat source, be aware that a thermometer cannot read the heat provided, so you will need to be led by the behaviour of your chicks. If they are always away from the end of the brooder where the heat source is situated and/or scattered around the edges, it is very likely too hot for them; if they are mostly huddled together at the heat source end or in a corner, it is probably too cold for them.

Do take time to get the temperature right. Being at the correct temperature is what drives chicks' natural behaviours, and promotes healthy eating and drinking. Safety is paramount, so make sure you observe the following:

- Use only EU/BS kitemarked equipment.

- Be aware of wires. Immediately replace any wiring that has become frayed or damaged. Keep wires out of water and inaccessible to the chicks. Don't suspend the lamp by the wire, use a chain or similar (lamps usually come with one supplied).

- Don't exceed the bulb wattage recommended for the lamp.

- Position the bulb away from anything that may melt or ignite – test the temperature of nearby surfaces with your hand after the lamp has been on for 10 minutes.

- Keep any water containers to the other end of the brooder – water splashes can shatter a hot bulb.

- If using gas as the source of heat, ensure ventilation without draughts and install a carbon monoxide detector.

- Position the heat source towards one end of the brooder, so that there are cooler areas for any hot chicks to move into.

Lighting Heat lamps are sufficient as light sources too, and it is one of the reasons I use them. Infrared or gas heat sources may require additional lighting unless the natural lighting is adequate.

During the last week or so before you move your chicks outside (see p.106), begin the transition to periods of darkness. Ideally this should be gradual, with natural light moving slowly from day to night, or with unnatural light mimicking this by part shading the light source. Try to avoid abrupt light/dark changes as this can startle your chicks. If possible, keep them in a room where lights need rarely to be turned on or off.

Bedding The term 'bedding' is slightly misleading, as the loose floor covering not only offers something to snuggle down into, but it also gives chicks' feet purchase as they learn to walk, and soaks up moisture and absorbs smells from droppings to some degree.

Softwood shavings are excellent bedding material and they are widely available. Newspaper makes a good base below the shavings, but it is too slippery when wet to use as the top layer, as chicks can't grip sufficiently when learning to walk. Avoid hardwood shavings, straw, hay and other grasses as they are unlikely to be sufficiently absorbent, they tend to get mouldy quickly and they may well hold traces of chemicals.

Change the bedding when it looks wet and if it is smelly. How frequently you need to change it will vary depending on the number, breed, age and stocking rate of your chicks. It is usually much less stressful for your chicks if you take them out of the brooder and put them in a large box while you change their bedding. Do this regularly and there should be no need to use a disinfectant.

When the chicks leave the brooder, clear out all bedding immediately. Wash the brooder with a good disinfectant and rinse well, leaving it ready to use when you have another batch of chicks.

Food and water for chicks

Newly hatched chicks can survive perfectly happily for a day or two without being fed – they have some of the egg yolk on board to see them through. However, you should be prepared well before they punch through the shell, with food and water – and the containers you'll serve them in – ready before they are needed.

The best food for chicks is a starter feed (usually made up of grains) or chick crumb (a fairly fine compound feed). Both are widely available online (see the Directory, p.246), in pet shops and country suppliers, but do check and/or order ahead as there are seasonal peaks and troughs in demand.

Medicated feed – loaded with antibiotics and other medication – is also an option, but for the home chicken keeper it usually isn't necessary. The additions provide protection from parasites and diseases that are much less likely in domestic situations than in large commercial operations. They also have a withdrawal period, i.e. a period of time after feeding where you should not eat either the chicken or its eggs. This may prove tricky, given that meat birds can be ready for the table in just 3 or 4 months.

Chicks quickly develop the instinct to scratch about. They'll scuffle their bedding around and disperse the feed itself, so while a bowl will do if you're happy to keep clearing out and refilling it, it's less grief to start with a feeder that allows the chick to peck in through openings but not to climb in and scratch it about.

Chicks need a constant supply of water. A tall plastic drinker is ideal as it has an internal tower of water that lets out just enough to keep a moat full of water for the chicks to drink from. These are cheap and they work well. Keep the drinker (or whatever water source you use) away from a lamp to keep the water cool.

If you decide to use a more open container for water, keep it shallow (less than 4cm of water) as chicks can get wet (and cold) easily, and can even drown. It will also need cleaning and refreshing with new water frequently – at least a few times a day. Whatever drinker you choose, keep it very clean.

Chicks become inquisitive about the outside world and may use their drinker as a leg-up to help them out of their home, so buy one with a domed top if you can.

Chicks to adults

You may be surprised by how quickly your chicks develop. If they are warm and well fed it seems as though you can almost watch them grow.

Over the first weeks you'll see their downy covering give way to feathers – firstly as pin feathers at the wing tips, gradually spreading over the tail, back and neck.

When they are a month old, assuming they have a good covering of feathers and the weather is warm and dry, any brooder-raised chicks are usually ready to leave the brooder and move to an enclosure outside. The temperature shouldn't be allowed to drop below 10°C, day or night, for the first weeks outside – use the heat lamp if necessary.

The ideal way to house the young chickens is in a run and house within, or adjacent to, the main enclosure, so that the main flock (and the mother, if present) and newcomers become familiar with each other while removing the possibility of the chicks being attacked by the other birds. It also lowers the risk of predator attack and prevents other hens from 'adopting' any of the chicks. The minimum space your chicks should move into is 100 x 50cm for up to three, with an extra 30cm^2 for each additional chick. Keep up with the chick feed (and water) but begin to combine with a little grower pellet feed until you are confident all the chicks are eating the grower pellets. Increase both water and food rations as the chicks grow. Make sure there are perches for your birds to roost on, about 20cm above the ground – they will start to want to do this as you move them out of the brooder.

At 6 weeks old, your chicks may well shock you with their size. For some meat birds, this may well be halfway through their lives, and for most layers they are a third of the way along the road to laying their first eggs. All chicks should be eating grower feed now.

Chickens become adults at around 18–25 weeks old and this is the time to move them into their permanent home. You may notice young chickens starting to play out their adult roles a little early – sitting on imaginary eggs, arranging nesting material, etc. When chicks begin to lay, usually at around 18 weeks old, it indicates their sexual maturity: they are adults. By this stage you need to have their nest boxes ready for them to lay in, and to have changed them over to layers' feed – do this gradually over a few days to allow your chicks to get used to their new diet. Don't be surprised if the first eggs are small or irregular in shape – they'll come good in a week or so. Cockerels begin to crow when adult, so be aware that from this point they are capable of fertilising eggs – if you don't want this to happen, now is the time to separate cockerels from the hens.

You can begin to incorporate the new chickens into the main flock from this point, but do so gradually and when you have time to observe any conflicts that may take place. If their enclosure is near or within the main flock's, they will already have become used to each other to a degree but the more gradually this integration occurs the more likely it is to go without incident.

Planning for cockerels

You are likely to find that around half the chicks or eggs you start with turn out to be cockerels. This is fine if you fancy having more cockerels and/or you are raising them for meat and the sex is unimportant. Otherwise, you should make a plan for what to do with them well in advance of buying chicks or eggs.

If you are raising birds to lay eggs for the table then there are the options of rearing the cockerels for meat, keeping them as pets, selling or re-homing them as cockerels or culling them at birth. You may find you can pass on the odd cockerel, but don't assume you can – they can be noisy and the market for them is often limited. They may well become aggressive, especially with each other, so be clear about your plan for them before you part with any money. A flock of birds with an undecided fate is expensive and potentially unsettling: their wellbeing should be top of your list of priorities.

Caring for Your
Chickens

Most pets naturally enjoy a routine and chickens are no different. Go with it. Your joint routine makes them feel secure, ensures they get fed properly, gives you regular interaction with them and means that feeds are unlikely to be forgotten.

Setting aside regular times to feed and water your chickens, to change their bedding and give them a look over, means you pay attention to them frequently and this is the cornerstone of caring for your chickens. You can't pick up a change in a chicken's behaviour if you don't know what that chicken's 'normal' is.

Daily care

Top of the list of essential daily tasks is to let your chickens out of their house in the morning and lock them up again in the evening. This is more time critical than at first it might appear. Letting them out early enough means they get sufficient natural light to lay well and to promote health. Locking them up as near as possible to the time that they have climbed into their house helps keep the fox at bay.

Chickens would naturally hop out of their house at dawn, but rest easy; allowing the sun to get a little higher before you let them out helps protect them from any early-morning fox patrols, as well as giving you a little extra sleep. We let ours out an hour or so after dawn in the winter and somewhere between 6.30 and 7am in the summer. The odd post party lie-in once in a while isn't going to do any harm, but do be aware that some chickens are prepared to make their feelings known vocally if the door to their house is not opened early enough.

Chickens usually feed little and often and therefore appreciate a fairly constant availability. Some owners like to use a feeder that provides several days' worth, while others prefer the routine of morning and evening feeds. Either is fine, but if you decide to feed twice daily, make the second feed of the day an hour or two before sunset as chickens don't eat at night (unless they are the fast-growing meat birds raised in intensive commercial conditions). What is important is to give your birds the right feed for their age and purpose, i.e. whether they are being raised for eggs or meat. See pp.113–19 for more about types of feed.

Your chickens should always have plenty of clean water – they're likely to need much more than you might think, especially on hot summer days. An adult chicken can easily drink a litre in a day when the weather is warm and even a brief period without water in hot weather can cause serious stress to your birds. We use a water dispenser which has a central barrel that can be filled, acting like a reservoir as it refills the available water tray on the outside. That said, chickens aren't keen on stale and/or warm water, so do replace any leftover water frequently, especially in warm weather.

A hen emerging from the house

Food will bring your chickens together, making them rush from wherever they are in their enclosure, and at this time it is often easy to spot any birds having difficulty in moving or any bullying behaviour. Take time to examine how they are behaving and whether anything unusual is happening.

Feed times are also good for checking over the fencing and the house. Ensure everything is secure, look for any signs of a predator trying to get in, and check that the electricity is working if you're using that kind of fencing. At both feed times, collect the eggs. This discourages either hens or predators from eating them and means you get them into the kitchen at their freshest.

If you can, even if it's not every day, spend some time just being with your chickens. Even 10 minutes makes all the difference. Let them get used to your company – most will come to you and peck from a handful of grain. If you're always in a rush, the pleasure of simply seeing them going about their business will be lost on you.

Almost all birds take themselves off to the house at dusk and you'll quickly get used to the time to turn up and lock the door behind them. Don't leave this too long – foxes are clever creatures and they usually don't need a second invitation. Occasionally, a new arrival may need persuading into the house as night approaches – a trail of grain and/or a little shepherding is usually enough to get them in. Any stragglers soon get into the habit of going themselves.

Weekly care

Cleanliness is the first step to maintaining a happy healthy flock. Allow an hour a week to clean the house and any areas of the run that need it, and to change the bedding within the house. Take out the old bedding and all the poo – one of the most effective catalysers in a compost heap, this is gold for the gardener. Forked into the pile (rather than chucked on as a single layer), it will speed up and enrich the process of compost making. Even if you aren't a gardener, don't throw it away: pass it on to someone who'll be grateful of its magic – a neighbour, perhaps, or someone at a nearby allotment. One word of warning: don't be tempted to add it straight away to your garden; it is powerful stuff when fresh and, before it has begun to break down and mellow, it can damage plants.

A wallpaper scraper is perfect for removing stubborn deposits from the floor and perches. Wash everything using hot water and (if you need to) a disinfectant that is safe for poultry; for suppliers, see Directory (p.246). Using a dry disinfectant in winter helps ensure that the house isn't damp when the chickens return to it.

This cleaning timetable provides the perfect opportunity to carry out those checks that are easily overlooked. As well as the regular walk around the perimeter of the enclosure, check that the house doors and roofs in particular are in good condition and that you are aware of food levels and whether you need to buy more.

A weekly check over your birds is also a fine habit to get into. As well as inspecting them for pests and diseases such as lice and mites, this ensures you are familiar with your birds and more readily able to identify any potential problems early.

Lastly, check the droppings for worms – if you don't worm your chickens as a matter of routine, treat them at the first sign of trouble (see p.125).

Feeding

Wild, undomesticated chickens would naturally eat a mixed omnivorous diet of green plants, grains, small insects and worms. They're quite partial to a bit of meat when the opportunity arises. The domesticated chickens we keep for eggs and meat are, in evolutionary terms, a fair distance away from these wild birds, having been bred to be highly productive. To maintain that productivity they need feeding well, which is all about the composition, quality and timing of their food.

In order to develop and be productive, chickens need a balanced healthy diet. This is best provided with a good-quality 'compound feed' and access to vegetation. The make-up of a chicken's compound feed varies, depending on whether it is a laying or meat bird, but broadly speaking around a fifth of a chicken's diet should be protein. Free-ranging birds can obtain these amino acids from bugs, slugs and worms, but in most cases these aren't available in sufficient quantity and most of their daily protein intake comes from their feed.

Buying cheap, low-quality feed is a false economy, as it generally implies low protein levels. Hens need enough protein to produce good-quality eggs consistently and frequently and meat birds will develop slowly and less healthily with insufficient protein. It pays to spend a little more on good feed.

Chickens don't need much in the way of fats and carbohydrates, and most diets in which grain forms a part will easily provide enough to keep them happy and healthy. Surprisingly, vitamin C can be manufactured by the chicken itself, but all other necessary vitamins must come from its diet. Some will be acquired from free-ranging, but a good-quality feed will have the levels of vitamins that encourage health and vitality in a bird.

Choosing feed

There are a few different types of feed on the market and plenty of different brands. It can be a confusing array but it is fairly easily navigated if you are aware of what you need to know. Check the label to ascertain what the feed contains and the expiry date – you need to be sure you'll use all the feed in the bag before this is up.

Check the list of ingredients. It is particularly important to be sure that the protein percentage is appropriate for your chickens at their stage of life. Also look for genetically modified ingredients. We use organic GM-free pellets to be sure there are no GM elements to their feed and no pesticide or other harmful chemical residues. Also check whether the feed includes medication. It is unwise to give your chickens unnecessary chemicals, and after they've had medicated feed, a period of time must elapse before you can safely eat either their meat or eggs. As a rule, if your chickens have a good balanced diet, supplements are not needed.

Feed is available as pellets or mash, the latter being smaller. I prefer pellets, as I think the size suits the way a chicken eats, but many like mash, and nutritionally they are the same, so go with whichever you prefer. Grains, available as mixed corn or whole wheat, are also good as a regular treat.

We feed our layers on organic layers' pellets in the morning, make sure they have access to enough of it in their feeder through the day, and then give them a few handfuls of organic grain in the afternoon, with the occasional handful of greens after that. Along with the grass, insects and worms they scavenge through the day, this gives them great variety while ensuring they don't fill up on greens. Our chickens have constant access to grass but if yours don't, do supplement their pellet diet with greens – a past-its-best cabbage, a bolted lettuce or similar, hung up for them to peck at, provides them with as much entertainment as it does nutrients.

You may find a local supplier who will mix feed to your own blend, if you are able to tell them what your chickens require. However, this is rarely a good option, as most set a minimum weight that is impractical for the domestic keeper and you are likely to find the price fluctuates considerably, depending on the current cost of each ingredient. To get a rough idea of how much feed you'll get through, allow an average of 125g per bird per day.

Chickens will snaffle up almost anything you offer them, but it is illegal to feed them anything from your kitchen. A cabbage from the veg patch is fine; one from your fridge is not. This is to minimise the possibility of notifiable diseases being circulated through the food chain, largely through cross-contamination with meat in your kitchen.

If you are planning to alter your chickens' diet, even from one brand of feed to another, do so gradually, adding half a bag of the new brand of pellet to the last half-bag of the old. Chickens can be fussy and may even go off their food if it is switched suddenly.

Chick feed

Chicks cannot deal with adult food. They require a chick crumb that comes in a particle size and nutrient mix to suit them. This will contain 20 per cent protein. Chicks from quick-growing commercial meat-bird breeds should be fed a higher

A brassica stalk, pecked of leaves

protein feed (22–24 per cent protein), usually sold as 'meat bird starter', but most slow-growing meat birds will be fine with the 20 per cent protein feed.

Pullet grower feed and layer feed

Pullets are chicks that have yet to lay. Once past the first month of life on chick feed, layers should move to a pullet grower feed. Typically this feed has a lower protein content (of around 18 per cent) to allow the birds to grow more slowly, enabling them to develop the legs and bone structure they need to be able to lay well. As with changing feed brand, make the last week's feed a mix of chick crumb and growers' pellets. Similarly, when reaching the point of lay at 18 weeks old, give them a week of half growers' and half layers' pellets, mixed together, as they make the change to 100 per cent layers' pellets.

Hens reach maturity at 18 weeks or so and begin to lay. At this point their diet changes for the last time, dropping a little in protein content to 16–18 per cent and increasing in some nutrients, such as calcium.

Meat birds' feed

After their first month or so on chick feed, most meat birds move on to a grower feed containing 18 per cent protein content. Then, at 12 weeks or so, they move to a lower-protein feed (16 per cent) until they are at the required weight, usually around 20 weeks or so. There are some breeds (Ross Cobb hybrids, for example) that have been selected to grow extremely rapidly – often ready for supermarket shelves at little over a month old – and they require very particular management. In truth, these birds are little fun to raise and lead a short, not especially happy, life that requires expertise to oversee. I suggest you leave them alone.

Grit and oyster shell

Because chickens have no teeth, they will often pick up enough small stones and the like, which they store in their gizzard to help break up food. Providing them with a little grit supplements any small stones they pick up, helping to ensure their food gets ground up before digestion.

Crushed oyster shell is also a fantastic source of calcium, which chickens need in reasonable quantities for healthy bones and for repeated formation of eggshells.

Both grit and oyster shell are cheap and should be made available to the chickens to peck at as they need it – they won't overfeed on either.

Poultry spice

Not the forgotten member of the all-girl band, but a mineral supplement that can be useful if any of your birds are recovering from illness, laying soft-shelled eggs or experiencing a heavy moult. There is no need for supplements other than this.

Mixed grain for laying hens

Occasional treats

Adding a little variety to your chickens' regular diet will be welcomed. Non-citrus fruit and vegetables, providing they haven't been sprayed, make nutritious treats once in a while, as well as providing something to play with and fight over.

What not to feed

Chickens are not necessarily the most discerning of creatures when it comes to recognising what is edible. Flakes of rusting metal or peeling paint, slug pellets and a jar of white spirit are all likely to be investigated, so try to be aware of everything they have access to. Potentially harmful chemicals in weedkillers and pesticides can be passed to chickens either by direct consumption and inhalation or by your birds eating the organisms that have ingested the chemicals. Obviously the best way to avoid potential problems is not to use chemical fertilisers, pesticides and herbicides at all. But if you do use them, ensure there is no possibility of contact with your chickens or their forage.

THINGS TO AVOID	REASONS
Scraps from the kitchen	They may include traces of meat
Milk and dairy	It is against the law to feed them either
Bread, breakfast cereal, cake and similar starchy foods	They can lead to compacted crop
Meat of any kind	Not suitable for domesticated chickens, though what they forage for is fine
Any other animal feed	It has the wrong composition for them and may do them considerable harm
Anything mouldy	Some moulds can be harmful to chickens
Leaves of tomato, potato, pepper, aubergine or rhubarb plants	These are poisonous to chickens
Avocado	It contains a chemical (persin) that can kill birds
Datura, nicotiana, sweet peas or morning glory	These plants are poisonous to them
Raw peanuts or beans	They can contain toxins

How much food

There is no fixed correct amount of food to supply your chickens – it varies with breed, size, time of year and many other factors. The simplest method is to make food constantly available through a feeder that stores more than they require each day, so it should never become empty. Alternatively, you can feed your chickens in the morning and adjust the amount each time you feed them until only a little is left by the time you come to feed them again. Meat birds need constant access to food, so ensure it never runs out.

If your chickens leave food on the ground or in the feeder it is a sign that you are giving them too much. Not only is this expensive, it encourages pests such as rats to help themselves at your expense. Similarly, if you notice any of your birds are on the thin side, it may be due to an ailment, but it could be due to insufficient food – if they are low down in the pecking order, there may simply not be enough left for them to eat. It can help if you set up two or more places to distribute the food, splitting up the flock and so reducing any bullying behaviour.

Catching a chicken

At some point, be it to transport a sick bird, to dispatch it, tame it or to move it easily from A to B, you will need to catch a chicken. Contact with your chickens is also part of the fun of keeping them, and learning how to do it safely and in a way that causes your birds least discomfort and stress is one of the first things on the 'to do' list when you decide to keep chickens.

It is all too easy to suffer humiliation in the pursuit of an uncooperative chicken. I can testify that the combination of a short, nippy and scared creature with a tall, increasingly inflexible pursuer is a recipe for a disaster.

Don't try to sneak up on a chicken and leap at it. In that once in a million time that it works, you will probably squash and harm the chicken.

Whenever possible, wait until night-time. Your birds will be at their sleepiest and it should be quite easy to lift the lid to your chicken house and take the bird from its perch. If you have a walk-in chicken house, use a blue bulb or blue filter in your torch as chickens can't see in blue light. If possible, use a head torch to leave both hands free to deal with your bird. Pick it up and carry it cradled under one arm (see p.122).

However you do it, have a plan, work quickly but not rashly, so as to keep disturbance to a minimum.

If you have to catch a chicken in the daylight, firstly find a way to enclose it in as small an area as practical. A bird's instinct is to be aware of attack from above, so some people go for the legs. A swift move to grasp the legs, while simultaneously

Netting a chicken

taking the bird's body weight, can often prove successful without inducing panic in the bird. Others feel that this tactic risks injury to the bird and prefer to use a little food to entice it, then press down firmly but carefully on the bird's shoulders, causing it to crouch so that it can be picked up easily.

A net can work well but calm, fast action is required. Before attempting to net the bird, throw down some food to (hopefully) distract it. Make sure the mesh of the netting is fine and use a sideways scoop rather than come down on the chicken from above.

A catching stick, which acts like a crook around the bird's feet, is used by some but I've never liked the idea of them – having the speed and confidence to hook the bird's legs without risking injury is a difficult balance to strike.

Chicks are relatively easy to catch. A hand in front of the bird acts as the bucket to gently but swiftly scoop the bird into with your other hand.

Retrieving escapees

Chickens like to escape, but in general they also know what's good for them – in other words, where the food is coming from. They also like to roost somewhere familiar, so their urge to return will strengthen as the sun goes down.

When (rather than if) some of your chickens escape, don't panic. Your first job is to ensure that once back in they cannot escape by the same route – throw a little feed down to keep them nearby if necessary, repair the netting, make good the fence, etc, before you try to retrieve them.

Leave them a clear route back into their enclosure (or out of the neighbour's garden if they've escaped from your property) and make a fuss of feeding time, so it's obvious food is being served. Any birds that remain out at dusk can be watched to see where they roost. If you can reach them, they'll usually be easy to pick up and return to their house. If any aren't in by nightfall, again don't panic – chickens are essentially sociable creatures and the likelihood is they'll wander around the outside of their enclosure trying to get back in when you go out the next morning.

Handling and contact

However anxious you may be of handling a bird, try not to be afraid of the situation or tentative in your actions – holding the bird at arm's length or too gently can invite panic and attempts to escape. Similarly, don't be too macho, and certainly don't hold a chicken upside down by its feet as a matter of routine.

Cockerels can be a particular challenge to hold. Wear long sleeves to protect yourself against any sharp spurs and use a hood or small towel to cover its head – this usually calms it down.

Hold a chicken by cradling it under one arm, against your body, with its head facing backwards – a firm cuddle is the sort of thing, rather than a full-on squeeze or a limp handshake. You can use your other hand to hold its feet if necessary.

Chicks need to be held with a firm but loose hand. If this sounds confusing, think of your fingers grasping an imaginary tennis ball – your fingers are fixed in place but there is no pressure on the space where the tennis ball would be. This is important, as chicks are easily suffocated by being held too tightly.

Handling your chickens is also the point at which diseases, bacteria and pests can be most easily passed either way. To ensure the risks are minimised, be aware of the following hygiene considerations:

• If you have visited another flock of chickens (or turkeys, geese or any other poultry), wash your hands well, change clothes and wash or change footwear.

• Wash your hands well every time you deal with your own chickens.

• Use rubber/latex disposable gloves when dealing with ill, possibly ailing or dead birds.

• Quarantine any birds you know or suspect are ill.

• Keep chickens away from your face.

Promoting health

Keeping pests and diseases at bay is all about creating and maintaining a healthy environment for your birds. Ensuring they have fresh air, sensible stocking rates and a new patch of ground regularly are the foundations of keeping a healthy flock. Their house and any equipment, including feed bins and water dispensers, should be cleaned regularly. It is repetition of the simple practices that prevents the majority of problems. Be observant: look long enough and often enough and you will notice anything unusual about your chickens.

Entertaining as they may be to watch, dust baths can be a real nuisance, located as they often are in your flower or vegetable beds. But they are one of the best ways of keeping external parasites to a minimum. Ensure your birds have access to patches of earth where they can hunker down, casting dust onto and into their feathers, irritating any lice and mites and removing many of them in the process. If you want to keep your chickens off your garden, dig up the odd patch nearer their house for them to use instead.

Holding a chicken correctly

As a matter of habit, once a month or so, we put a peeled clove of garlic into the chickens' water. The flavour permeates the water and acts as a deterrent to many parasites; it also seems to give the chickens a general boost.

Once a month or so, we add cider vinegar to their drinking water, at a dilution of 1:20 cider to water, to flush the gut in the event that one or more of the chickens has mild diarrhoea. The water must be in a plastic container, rather than a metal one, as the vinegar can react with metal.

Giving your birds access to grit (soluble and insoluble) helps prevent sour crop (see p.146); they'll take it as and when they need it.

Rodents can be a particular nuisance – eating chicken feed, occasionally taking eggs and young chicks, and transferring disease. To minimise this potential, store feed in a secure bin (see p.48), clean up any spillages immediately and keep an eye out for any rodent droppings.

Vaccinations

Whether to vaccinate your birds, and against which potential diseases, is a matter for the individual to assess. It is largely about balancing the risk, the scale at which you are keeping chickens and the cost involved. For more about diseases and pests, see pp.152–6.

Considerations when deciding whether or not to vaccinate your birds:

- Vaccines are usually sold in packs that are larger than required by domestic chicken keepers, so expense and waste/storage issues can be high.

- Vaccinating only part of the flock is unwise – the vaccinated chickens can still carry and transmit the disease to the unvaccinated birds.

- Do you intend to show, breed birds or sell live birds? If so, they are likely to come into contact with birds from outside your flock and there is more reason to vaccinate them.

- If you intend to keep, say, more than 12 birds, or intend to significantly increase the size of your flock, seek advice from your vet.

- Your vet can advise you whether any diseases are common in your area. There are localised hot spots of each, and that may affect your decision.

- Marek's disease, mycoplasma, infectious bronchitis and gumboro disease (see pp.152–3) can be vaccinated against.

Generally, the smaller and the more closed the flock, the less risk there is of contracting diseases and the less need there is to vaccinate. Like many small-scale domestic chicken keepers, I choose not to vaccinate my birds. I don't show them, nor do I sell any chicks, so the risk of infection to my birds is small. And the fact that the flock is essentially out of contact with other birds means there is no risk of disease being passed on. If any of those factors was to change, I would reassess whether to vaccinate against at least some diseases. Indeed, I could be required to vaccinate by some show organisers, or those I might be hoping to sell to.

Worming

Unfortunately, worms, of the parasitic kind, love chickens. Roundworms, caecal worms and hairworms are the most common, but given half an opportunity your chickens will acquire any number of other types. Keep chickens and at some point it is fairly likely that they in turn are keeping worms.

As ever, prevention is the best course. Understocking (i.e. giving them more room than they need) and moving your chickens to new ground every month or two dramatically reduces the likelihood of worms. Otherwise, you should consider worming your birds twice a year, in spring and autumn (see p.149).

Most medicines come in a suitable form to add to the chickens' water or to their feed and are easy to administer. Wait until your birds are at least 18 weeks old before you worm them.

If you feel confident that your preventative measures are working well, you may want to consider the occasional application of one of the herbal worming treatments that are also available – many appear to work well and, in contrast to other wormers, do not require a withdrawal period so the eggs stay safe to eat.

Quarantine

There are times when you'll need to keep a bird or group of birds separate from the main flock. The most common reasons are illness, behavioural issues, birds returning from a show, or when you are introducing new birds to the main flock. Don't wait for the time to arrive: be set up ahead, at least in terms of knowing what you are going to provide and where everything you need is located.

If at all possible, locate your quarantine area in sight of the main flock as this reassures established flock members who are in quarantine and allows new birds to become familiar with the existing flock and vice versa.

Discouraging and dealing with bullying

A certain amount of scrapping is almost inevitable when keeping chickens – it is how they establish and maintain a hierarchy within the flock. It can occasionally morph into bullying and you need to have a strategy to turn to if it occurs.

Suggestions for avoiding and dealing with bullying:

- Bullying is often a sign that your birds are enclosed in too small a space – extend their range, increase their access to free-ranging and/or consider whether their housing should be extended.

- Enrich their enclosure with logs or something similar, as this enables some to hide and others to display their superiority.

- Ensure your chickens have sufficient access to a good balanced diet – while greenery is vital, protein is important too and this should be provided in their bought feed if they haven't access through adequate free-ranging time.

- Stress can lead to bullying, so avoid excessive handling and minimise loud noise. Be aware of predator presence, such as buzzards overhead. Also consider changing brooder bulbs to infrared as this can reduce stress.

- Be prepared to move the house/enclosure to change their surroundings – this calms bullying more often than not.

- Be aware that birds with colouring, feather patterns or particular markings that differ from the majority of the flock may be the first to be singled out.

- Act quickly where blood is drawn. Blood can lead to extended pecking and sometimes frenzied attack from the other birds, causing serious injury, death and even sometimes to the dead bird being eaten.

- If all else fails, be prepared to remove the bully or bullies, either for a period of quarantine or permanently.

Dealing with aggressive cockerels

Although often calm in demeanour, cockerels can turn aggressive almost overnight and rarely regain their former easygoing nature. Often sizeable, and possessing a sharp beak, toes and spurs, a cockerel in the mood to injure is a handful for an adult, never mind a small child.

The likelihood of an aggressive cockerel can be greatly reduced if you raise the bird from young, keeping contact with him – picking him up and handling him occasionally as he grows.

If a cockerel starts to become aggressive, you may be able to override his urge to dominate you. Wearing gloves and long sleeves, approach him and press down

A Copper Maran cockerel

Where to trim wing feathers

on his shoulders, causing him to hunch. Then pick him up, placing him under your arm, facing backwards. This will often diffuse his aggression, and if repeated a few times over a week or two can sometimes do away with his need to attack.

Often though, an antagonistic cockerel remains aggressive and you have two choices: manage the danger by being careful at feeding times and keeping children away, or dispatch the irritating devil (see p.160–7). We inherited a cockerel, Henry, from friends and although he was never the most easygoing bird, he became vicious overnight. We managed the situation for a while, tried different approaches for a few weeks but he was taking the pleasure out of keeping the chickens. He took very well to a long slow cooking in red wine, a cidery version of which is on p.201.

Trimming feathers

Chickens are hardly the Louis Blériots of the avian world. At a push they can flap themselves away from the earth for a few brief moments, but this can be enough to take them over a fence. If any of your birds show a tendency to do this you may want to trim their wing feathers. You only need trim one wing, as this unbalances the bird enough to make lift-off tricky, and not a little comical.

You only need to trim the large feathers used in flight (as shown opposite). It is possible to do this alone, but it is easier if you have someone holding the bird for you. Hold out a wing and cut across the flight feathers using strong, sharp scissors. There will be no bleeding and the bird will not experience any pain.

If you catch a young feather it may bleed. If this happens, grip the feather close to the body and pull it out. Although a little painful for the chicken, this stops bleeding and prevents the likelihood of other chickens pecking at the blood.

You will need to trim the wings again after the moulting period, when the bird regrows new feathers.

Breeding

Keep a few chickens and sooner or later you'll ask yourself the question: do I want to breed some of my own? It may be that you're intent on moving up the scale of productivity, that you want to provide yourself with the next generation of birds rather than buy them, or perhaps you fancy giving it a try for the sheer pleasure of it.

Firstly, think about what you will do with any cockerels that are born. If you are raising birds for meat, cockerels may be as suitable as hens, but if you are raising chicks to be layers they are not so good. Unless you are extraordinarily fortunate, around 50 per cent of the eggs will hatch into male chicks and you should decide what you plan to do with them well in advance of starting to breed any chickens (see p.107).

The next step is to decide how many birds you are hoping to raise. Consider how long the offspring will be with you and how much space you have available, then keep within the limitations of these constraints.

You can hatch your chicks either by using a hen (see below) or by using an incubator (as described in Starting with Eggs, p.93).

If you are raising birds for meat, the sex of the offspring is usually unimportant, so simply decide on the number of adult chickens you can look after and place the same number of eggs (plus one or two spares) under the hen or in the incubator. If you are breeding for layers, cockerels are of no use to you, and given that half of your eggs are likely to hatch out males, you can safely add extra eggs to give you a better chance of hatching the right number of hens.

Choosing what to breed

Most people who breed chickens for their own flock rather than for competition decide to produce hybrids. A hybrid results from parents of two purebred chickens of different breeds, or from a purebred parent and a hybrid parent. You can also get hybrid chicks from two hybrid parents, but the results are usually highly unpredictable.

Producing hybrids is a reasonably determinable, if inexact, science. Like mixing blue and red paint together, you know you'll end up with purple, but it is not easy to be sure of the exact shade. Generally speaking, crossing purebreds as parents results in more predictable characteristics in the offspring than if one of the parents is a hybrid.

Consider the qualities you want in the young. Chosen with care, some breeding combinations result in positive effects, such as faster growth, improved health or a greater laying rate. This is usually a result of exaggeration – for example, choosing parents from good meat breeds generally leads to offspring with similar, often more pronounced, traits. The principle also applies to layers. By the same token, a cockerel from a good meat breed crossed with a productive laying hen may well result in a good dual-purpose bird.

Beware of breeding from two hybrid parents: the complexities of genetics mean that the offspring are likely to have characteristics from their grandparents as well as their parents, and possibly variations on those characteristics. Feel free to try if you fancy, but don't expect predictable results.

There is nothing to stop you choosing purebred parents from the same breed. Chances are the offspring will develop many of the characteristics of their parents; in other words, they will be true to their breed. Some chicken keepers do this for the purpose of showing their birds competitively. This is a complex business, requiring a familiarity with the competitive standards for that breed. The Poultry Club of Great Britain (see Directory, p.246) and local groups should be able to advise you about this.

Parents and mating strategies

As well as choosing the breeds you want as parents, the choice of the individual birds is crucial. It is simple: choose your best cockerel and your best hen(s) from the breeds you want and you are likely to have the healthiest offspring.

Make sure the parents are fit and free from genetic defects and avoid choosing closely related parents. Consider the particular traits of individuals too. Some hens from a particular breed lay better than others, and there's a fair chance that this ability will be passed on to the offspring.

Good, healthy chicks come from parents that are in peak condition. Choose birds that are full of vitality and check them over thoroughly for any ailments or parasites. The older either parent is, the more important their condition.

Your chosen cockerel must be fully grown. Immature cockerels are sometimes dominated by females and in extreme cases may become impotent as a consequence of losing their position as leader of the pack.

There are two main approaches to putting the cockerel with the hen(s): flock mating, and pair/trio mating.

Flock mating

The chosen cockerel is allowed access to the flock in the hope of producing good numbers of fertile eggs within a short space of time. The disadvantage of flock mating is that it is hard to be sure which hens the cockerel is or isn't mating with and consequently you may not be able to tell which are the most fertile mothers. The optimum number of hens to each cockerel for flock mating is around six, as the larger the flock, the less frequently he will mate with each one. Around 14 hens is the typical threshold past which egg fertility is likely to drop significantly.

Pair and trio mating

This involves isolating the cockerel with one or two hens. The advantage of this method is that you can choose your very best birds, the cockerel will mate with the hen(s) regularly, which means a likelihood of more of the eggs being fertile, and it is easier to be sure of which bird is responsible for the best results in the offspring.

Use netting or a separate run to isolate your birds, and keep them separate for as long as you need to collect the number of (hopefully fertile) eggs you require.

Preparing for breeding

Feed quality is crucial in the run-up to breeding. Ensure the feed you provide has a good balance of nutrients, particularly calcium for egg production, and make some crushed oyster shell available to them. It is quality not quantity that is important. Overfeeding can lead to fat birds, which may have difficulty breeding. Examine the breast: if you can't feel the breastbone, it is likely that your bird is overweight, so cut back the rations; if it is too prominent, increase their feed.

Timing is important. During the colder end of the year, when the days are shorter, the fertility of both parents tends to be at its lowest and breeding is tricky. If you can artificially provide around 15 hours of daylight each day and keep the temperature above 5°C, you may have a good chance of successful breeding; otherwise, wait for longer, warmer days.

Bring the females to the cockerel – letting him stay on his patch is good for his confidence. Be aware that the cockerel's spurs may cause injury during mating. If your cockerel is suitably tame, you can soften the spur tips with oil and trim them, or buy a poultry saddle (see Directory, p.246), to protect the female from injury.

After mating, it is around 3 days before an egg may become fertile, so allow the hen to lay a few before you begin collecting, or leaving the eggs for hatching. The sperm is stored in the infundibulum (where fertilisation takes place) and can remain viable there for a fortnight or so. This means you should isolate the hen(s) from other cockerels for a fortnight before you start collecting or leaving eggs for hatching, so you can be sure that the intended cockerel is the father.

Eggs chosen for sitting or incubating indoors should be clean, less than 10 days old and free from cracks. Other than that, feel free to choose whichever eggs you like the look of, remembering to keep to the number of chickens that you can accommodate and look after for the duration of the time they are with you. The rest of the eggs can be eaten, within the usual constraints.

A Copper Maran cock and partner, isolated

Generally around 75 per cent or so of the fertilised eggs will hatch successfully but rates are variable and 100 per cent is not unheard of, so never try to hatch more than you can look after.

Fertile eggs stay in a 'potential' state and the embryo won't begin developing until the warmth of a hen or an incubator is felt. For advice on using an incubator, see pp.93–4; for information on using a brooder, see pp.102–4.

Hatching using hens

A broody hen is nature's gift to anyone wanting to breed chickens. Instinct compels the hen to keep the eggs at the correct temperature, to turn them frequently, to help them hatch and then be a 'mother' to the developing chicks. Most broody hens will look after eggs from other hens; she may well take duck or goose eggs too.

Hybrid birds rarely go broody, whereas purebreds usually do to some degree. Most birds that go broody do so perhaps three times a year. Crossing a bird that is prone to broodiness with breeds that are prone to frequent broodiness can magnify this trait in the offspring.

You will need to provide the mother and newly hatched young with a separate, safe, enclosed space, with their own house and food and water. As well as keeping predators out, this keeps the chicks away from the attentions of other hens. A small enclosed run with housing attached is perfect.

Bring your broody hen, the eggs and the house you intend to keep them in together at night when the hen is at her sleepiest. Place the eggs underneath your hen, ensuring they are of approximately the same hatch date – any added from a later batch are likely to fail as the hen will leave the nest to care for the newly hatched first batch. Check the next morning that she is still sitting on the eggs and that they are all in place; otherwise, leave her undisturbed.

Your hen may sit almost all day, leave the nest only once or twice to eat, drink and poo. This is perfectly normal. If you suspect that she is never leaving, as occasionally happens, you'll need to gently persuade her out of the house once or twice a day.

You may notice that the hen has kicked an egg away from the rest – this could be accidental so try to return it to the rest. If she kicks the same one away again it is possible that the egg is infertile – hens are often good at telling when an egg is not developing properly.

If you want to be sure that embryos are developing as they should, you can check by candling (see p.95) but do this only once, at night, so as to minimise disturbance. If you are in any doubt as to whether the embryo is present, I would suggest leaving that egg with the mother, as there is little to be lost by doing so.

As the time for hatching approaches at 3 weeks, the hen may leave the nest even less. She'll be able to tell from the slight movements within the egg when hatching is imminent. At this point, leave food and water by her in the nest rather than insist she leaves to eat and poo. Don't disturb her or the eggs unnecessarily. Listen for cheeping from chicks newly emerged or still in the egg, or the sound of eggs being chipped by those starting their journey into the outside world. Only if you feel something is wrong should you lift the hen to check – and even then be careful as eggs are easily lifted along with the hen and can fall and smash.

It is critical that you let chickens extricate themselves from their shell. Even in circumstances where the rest have hatched and there seems to be no sign of progress, most breeders will advise you to leave chicks alone. I am aware of some experienced breeders intervening once in a while to help a late chick out but mortality rates are high.

Usually 40 hours or so after the chicks have been born, the hen will leave the nest with them in search of food and water. Have chick crumb and water ready well in time for this. The hen will teach the young how to eat and drink.

One last note: deformities at birth do occur, some of which are covered in Health Issues and Predators (see pp.138–57). Occasionally the hen realises that something is wrong and kills the chick; otherwise it may fall upon you to take it to the vet or to do the deed yourself. Breaking its neck or using dispatch pliers are the quickest, most reliable methods.

A Wyandotte cock with hens

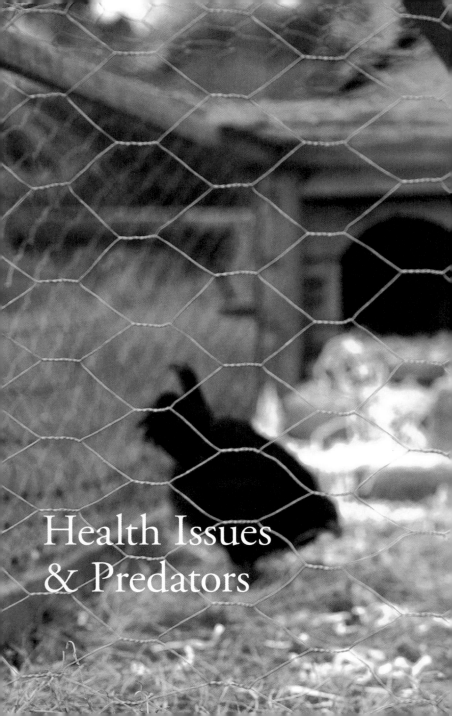

Health Issues
& Predators

Chickens are pretty hardy creatures but, like all animals, they can get ill from time to time. Inevitably a weak bird is more vulnerable to attack and chickens have evolved a strategy to disguise the fact that they are carrying an illness, so spotting an ailing chicken can be tricky. Unfortunately, the first sign may well be its feet sticking up in the air.

Chickens can keel over having reached a natural end, or owing to a genetic problem, but death may also be the result of disease or pests, so you need to become familiar with how to examine a bird. You should also learn to recognise a few of the potential nuisances that may affect them and know how to react if injury occurs, or pests or disease strike.

Don't be put off by this chapter. Serious ailments are rare on the domestic scale, but the occasional troublesome appearance of something mild is almost inevitable. It bears restating that habit and observation are the cornerstones of keeping healthy chickens – clean them out often and regularly, keep their diet and surroundings healthy and varied, and take time to observe them. Follow sensible precautions: wash your hands and footwear if you have been to visit other flocks, and introduce new chickens to the flock only after a period of quarantine (see p.125). Also make sure you are registered with a vet before you need one.

Your health

As long as basic health and hygiene procedures are followed, you are unlikely to come down with anything other than a cold caused by locking your chickens up on a cold, windy, wet evening in winter. Although particular strains of avian flu are serious, outbreaks are extremely rare and the chances of acquiring it are minimal. Salmonella infection (see p.153) is more of a potential threat, causing gastrointestinal problems that can be serious in the young, old or those with any underlying health issues. Thorough washing of hands following any contact with your chickens, their enclosure or any related equipment is usually enough to remove any risk.

Checking a bird over

You're very unlikely to determine exactly what is troubling a chicken by observing it from afar: you'll need to pick it up and examine it. Follow the guide to catching a chicken (pp.119–21), minimising any stress. In your understandable haste to care for your bird, don't be tempted to chase it, as this can greatly raise its stress levels and increase any bleeding that is occurring. You may find an injured or unwell chicken seeks the house – it will be much easier to get hold of to examine if it does.

Examining a bird

Gentle firmness and confidence are key. Don't panic. Wear gloves and have a bucket of warm water and a sponge to hand. Part the feathers and look for wounds or pests, whichever you suspect.

If it is a wound, take time to wipe off any bleeding areas – you need to get an idea of the extent of any injuries, and puncture wounds can often be deep rather than wide. If wounds are large and/or numerous you may be faced with a bird that has no chance of survival. Hard though it may be, the kindest thing could be to end its life (see p.160).

If it is a pest or disease you are hoping to identify, examine the chicken carefully for any signs. Make a written or mental note of any symptoms, noting where they are found, how extensive they are and any other details. The more detailed the information you have to hand, the more accurate the diagnosis and treatment is likely to be.

Wounds

Left alone, most minor wounds will heal by themselves fairly quickly, but this rarely happens with chickens. Others in the flock become fascinated and their unpleasant instinct to peck at injured birds kicks in. This can make small wounds much worse in a short space of time. Applying a coloured antiseptic spray can work wonders, helping to heal the injury and, equally importantly, disguising it from other birds. If you are at all concerned, isolate the affected chicken and be prepared to call the vet if required.

A chicken's skin is easily damaged and serious wounds often result from attacks by dogs and other predators. If a chicken has wounds of this sort, you should:

• Clean the wound(s) well using mildly soapy water.

• Flush wounds well with salt water (2 tsp salt to 1 litre of warm water), iodine or tea tree oil to disinfect.

• For any wounds that bleed profusely, use a wound powder – available from most pet shops and vets, or see the Directory (p.246) for suppliers.

• Check the wound a few times a day for any signs of infection.

• Clean the wound as suggested three times a day if infection is present, covering with a wound dressing (available from pet shops, vets, etc) to prevent flies laying eggs in the wound.

If your bird is more seriously injured but you believe it may live, you'll need to take immediate action. If you decide to take it to the vet (which can be expensive), call ahead to check how long you will have to wait. The presence of other (often noisy) animals in the waiting room for longer than is necessary may well add to the stress of the injured bird.

Whether you decide to home-treat your chicken or take it to the vet, you will need to create a comfortable, clean, safe area, away from the other birds. Warmth, semi-darkness and water are the priorities. Use a heat lamp, or similar, to provide warmth if the weather is cold.

Whether home- or vet-treated, if your bird doesn't begin to improve reasonably quickly following treatment it may be that it has sustained wounds that are too serious to recover from. You should be prepared to carry out a humane way of dispatching your bird (see pp.160–7) to avoid a slow death.

Injuries and other ailments

If a chicken has broken a wing it may not be as serious as it looks. Carefully fold the wing back into the usual position and use wrap (available from the vet) to stabilise and secure it. Quarantine the bird and allow it to recover for at least a fortnight. If the break is a bad one and/or bones are showing through the skin you should consult a vet. Amputation is sometimes an option, as birds can often survive happily with a single wing.

Broken legs are more problematic. Load-bearing as they are, legs take time to recover and although a bird can survive with a single leg it is not usually a happy existence. If you suspect that a chicken has a broken leg (it may be crooked and/or swollen), you should consult your vet. The affected limb can often be splinted and will heal quickly; keep your bird quarantined while it recovers. As ever, if the bird's prospects are judged to be poor, dispatching may be the kindest thing (see p.160).

A chicken's comb and wattle are easily damaged in fights, or if they catch on something sharp. Often these wounds will heal naturally and you should use an antiseptic, such as tea tree oil, to help the process and discourage flies. More serious injuries might not heal and trimming may be required. A vet can do this for you or you can do it yourself. Be aware that there are minor blood vessels in both comb and wattle and the chicken is likely to experience some relatively minor pain, but this is over very quickly and recovery is usually rapid and complete. If you are in any doubt about how or whether to trim, consult your vet. If you are doing it yourself, wash the affected area well with hot soapy water and clean it with alcohol. Trim off the affected area with sharp, sterilised scissors and apply antiseptic cream and an anti-fly cream. Quarantine the bird until the wound is fully healed.

Eye injuries are relatively uncommon, but a predator attack, fight or occasional snag on something sharp can result in damage. Use a non-medicated eyewash and cotton wool to clean the affected area. Isolate the chicken until it is fully recovered. A chicken can survive perfectly well with just one eye, but a blind bird will not be able to eat or avoid predators and should be dispatched (see p.160).

Chickens' beaks do not regrow, so any damage will need to be assessed. If the damage is limited and the bird can eat normally, then just keep an eye on it to ensure that it is able to get enough food and there is no decline. If the beak is more seriously damaged and the chicken can't eat properly it will need to be dispatched.

Egg binding Although this isn't overly common in home flocks, it is something to be aware of. It occurs when a chicken cannot pass an egg through the oviduct and may be caused by a genetic problem, by injury or when the bird is weakened by disease, pests or inadequate feed. An affected chicken will be lethargic, have ruffled feathers and spend her time sitting. Her tail may be twitching up and down as she tries to deliver the egg. You may even be able to feel where the egg is stuck but it is best not to attempt to move it as an egg broken inside your hen is likely to make the situation more serious. In most cases, there is only so much you can do to help an egg-bound chicken. The only safe course of action is to provide moist heat. I have, thankfully, never had to do this and the likelihood is neither will you, but here's what you should do if you suspect that your chicken is egg-bound:

- Create a wire-floored cage using a large cardboard box or similar.

- Place a large, wide dish or pan of steaming water beneath the cage.

- Provide drinking water for the chicken.

- Hang a heat lamp within the box to provide extra warmth – aim for a temperature of around 35°C.

- Put the chicken in the box, then cover the box with a large bin bag and/or a blanket to retain moisture and heat.

- Keep the water steaming by adding more or replacing it as required.

A couple of hours of this is usually all it takes for the egg to appear. If it does, allow your hen to rejoin the flock. If no egg appears but your chicken 'recovers' it is likely that something else is wrong, and you should consult a vet if you cannot identify what the ailment is. If no egg appears but your hen remains lethargic and unwell

you can continue the treatment for another couple of hours. If there is still no change, consult your vet as to the next course of action. Injections can sometimes help, but dispatch may well be the kindest thing.

Mudballs Thankfully this affects the feet of a chicken, where a combination of mud and bedding builds up to form a ball that makes walking very uncomfortable. As you'd expect, it is more common in wet weather. It is easy to damage a chicken's foot when trying to remove the mudball, so be cautious. Use plenty of warm water to soften any build-up and work patiently.

Bumblefoot This foot infection shows as a small red swelling on the underside of the foot, which can split, bleed, cause discomfort and even disfigurement. If left untreated, it can sometimes lead to death. Regularly check that all perches, flooring and steps into the house are smooth as splinters are a common cause. Do consult your vet if the infection is bad; antibiotics are the usual treatment.

Sour and impacted crop The chicken's crop is an internal store just beneath the neck where just-eaten food passes to soften a little before making its way into the stomach and gizzard to be fully digested. The crop is easy to see in young chicks, and with practice it is usually straightforward to identify in a larger chicken.

Problems can occur when some of the contents of the crop don't pass into the digestive system, instead fermenting in the crop, which can lead to infection. This is known as sour crop and is best checked for in early morning – the crop will feel watery and a sour smell will be obvious when opening the bird's beak. If the crop feels hard and firm, your chicken may have impacted crop, where the food has built up over time. If your bird has either ailment, it is likely to eat less, seem listless and make peculiar neck movements as a result of the discomfort.

To treat sour crop, hold the chicken upside down by the feet and gently massage the crop from bottom to top (i.e. towards its mouth) to encourage the sour liquid to be released through the bird's mouth. Only do this for about 20 seconds, or you may end up choking your bird. Feed it layers' pellets mixed with natural yoghurt and water to neutralise any bacterial build-up. You may need to follow this treatment for 3 days or so to resolve the problem. If it persists or you feel nervous of carrying out the procedure, speak to your vet, who can supply an anti-fungal treatment.

Impacted crop, where food builds up in the crop and hardens, is a little less tedious to treat than sour crop. A scant ½ tsp of olive oil, dropped into the chicken's mouth, followed by a gentle massage of the crop to loosen the obstruction should be undertaken morning and afternoon for 3 days. As with sour crop, feed with a mixture of layers' pellets and natural yoghurt. If the problem persists for longer than 3 days of treatment, consult your vet.

Providing poultry grit helps your chickens break down their food and minimises the likelihood of food building up in the crop. You should avoid treats of bread, pasta and similar starchy foods that promote compaction.

Feather pecking A little light pecking is all part of asserting and maintaining a hierarchy within the flock, but this can occasionally lead to bullying and more extreme behaviour. Adding new birds to a flock, overstocking (i.e. giving them too little space), vitamin deficiency and boredom are the most likely causes. You should act as soon as you see it happening as the problem can accelerate quickly, leading to feather loss and bleeding, which can sometimes trigger a frenzy that can even end in cannibalism. Isolate the aggressor and check over the victims, treating any areas that are bleeding with antiseptic.

Giving your chickens space, a few obstacles (such as an old tree trunk) to play around, and a mixed diet of layers' pellets, grain, greens and oyster shell helps to maintain healthy vitamin levels and will usually be sufficient to prevent bullying and excessive pecking.

A persistent offender can be fitted with a beak bit that allows them to eat and drink as normal but renders them incapable of feather pulling. Try it for a short time as it is often enough to break the cycle. If that doesn't work, you may want to consider preheating the oven.

Spraddled legs This is occasionally a problem for recently born chicks. Avoid keeping chicks on slippery floor surfaces as this often causes their legs to splay. If a chick is affected, lightly tie the chick's legs more closely (but loosely) together with string in a figure-of-eight and the problem should be rectified within 5 days or so.

Egg eating This is likely to happen at some point and may be due to a number of reasons. Overcrowding is the most likely cause, so make sure your chickens have enough space and that there are no more than three birds to a nest box. If they have the right space, check them over for general health – stress from worms, lice, etc, can cause this behaviour. Check for any chickens laying soft-shelled eggs as this can be an invitation to egg eating; if this is the case, treat appropriately (see p.69). Lastly, remember to take eggs away every day, as the more that build up in the house, the more likely the chickens are to start eating them.

If egg eating persists you need to break the habit by isolating the offender(s) and taking their eggs as soon after laying as possible, and/or filling a blown eggshell with mustard to surprise and dissuade them. To do this, use a pin to make a hole at either end of the egg and carefully blow the yolk out of the shell. Use a straw to suck up mustard (careful!), then blow it gently into the egg. A week or two of this should break the spell.

Parasites

Your chickens can offer a warm, comfortable home for a number of parasitic creatures. Providing a healthy environment for your birds is the most important thing you can do to minimise problems, but complete prevention is unlikely. Lice and mites may find their way onto the outside of your chickens, and worms and coccidia can affect them internally.

Knowing how to deal with parasites is crucial in preventing a presence turning into an infestation.

Worms Chickens can suffer all manner of digestive worms and while they are almost inevitable, there is much you can do to minimise the degree to which they become a problem. Moving the enclosure to fresh areas of grass makes a significant difference to the degree of worm build-up. A clove of garlic in the chickens' water and/or herbal preparations can act as a gentle means of minimising worm levels. However, these are inadequate when it comes to infestations and a vet's advice should always be sought. Only a few worming products are licensed for use with chickens and it is important to get dosage levels right and be aware of/observe any withdrawal periods.

Some like to worm their chickens twice a year, in spring and autumn, to avoid problems, but in most cases there should be little need to do so if you stick to the fundamentals of good chicken care. And if trouble strikes, remedy is usually swift and effective if you act quickly and worm them.

You're most likely to confirm that your chickens have intestinal worms by observing their droppings. Try to pick up the other signs before this: affected chickens tend to be thin, uncharacteristically unhealthy- or untidy-looking, their laying rates may drop and they may become weaker. It is important to be aware of such symptoms as most of us don't check chickens' dropping as a matter of course, and in some cases worms don't appear in the droppings. Your vet can do a worm count on a droppings sample if you are in any doubt.

Roundworms are tough little creatures, and although chickens can often survive happily with some worms present in their intestines, more serious infestations may seriously weaken or even kill. If you see a white worm up to 7cm or so in length in chicken droppings, it is likely to be a roundworm.

Tapeworms are fairly common, though not often harmful. They are picked up from any of the many intermediate hosts such as slugs and earthworms. Your chickens may become a little thinner and occasionally pieces of the tapeworm may be apparent in their droppings, but most owners are unaware that tapeworms are present in their birds. Usually their main damage is to your pocket – the chicken needs to eat more food to share with the parasite.

Chickens may also suffer from gapeworm, a lung-dwelling worm that causes the bird to stretch its neck out and open and close its beak, as if trying to swallow a rather long piece of bacon rind. This can be the most serious of the worm parasites because it may affect the chicken's breathing.

Lice and skin mites

Just writing that gets me itchy. One of the reasons chickens have dust baths is to rid themselves of lice and mites, but some can cling on. Check your birds regularly: part the feathers and watch for any small grey creatures darting about and/or tiny eggs at the base of the feathers near the skin.

Although not overly serious in the early stages, if mites are allowed to establish the birds will become stressed, weakened and they will decrease in productivity. A weakened bird is not only an unhappy one, it is susceptible to other disease, so act quickly. Powder treatments are available from many suppliers (see Directory, p.246). If the infestation is serious speak with your vet.

Red mite If you have chickens, sooner or later they will have red mite. These tiny mites can be hard to spot as they hide in nooks and crannies in the chicken house and pour out at night to suck the blood from your chickens, turning from grey to red as they do so. Get used to checking in the house at night once in a while and run some white kitchen paper along the perch – if you see red on it, it'll be squished blood-filled red mites. You may be able to detect them with a torch or even feel them on you. They can't live on or feed on humans, which is some comfort. Watch your birds for any signs of depleted energy. If they seem reluctant to enter the chicken house at dusk red mites may well have taken up residence. Left untreated, these mites can kill by slowly depleting a bird's strength, making them anaemic.

Infestations need treating with strong chemicals, so use preventative measures and early intervention. We use diatoms, a natural product made from powdered sedimentary rock, itself the product of fossilised remains of hard-shelled algae. The powder scratches the waxy shell of the mites, dehydrating and usually killing them. It also has some success in dealing with lice. Applied monthly to every corner of the house and on the birds' feathers at the first sign of trouble, it seems to keep infestations at bay and rapidly reduce numbers when required. Being a natural product, it has no withdrawal period (when you may not eat or sell the eggs).

You can run a blowtorch lightly around the inside of a wooden chicken house to burn off any mites or eggs that are hidden in crevices and corners. I do this three times a year (early spring, midsummer and early autumn) on a dry, sunny day and leave the doors and roof open to air until nearly sunset.

Whatever method you use, be thorough: red mite eggs can survive for years, waiting for optimum conditions to hatch.

Treating the house for red mite

Scaly leg mite This mite inflames the scales on the chicken's legs, making them swollen and painful. It is easy to spot once it has taken hold, but it is also a fast mover, so act immediately and treat the whole flock as soon as you spot it in its early stages. Treatments range from applying petroleum jelly to more powerful spray treatments, available from suppliers (see Directory, p.246) or your vet.

Diseases

Diseases that affect chickens can be very serious but are mercifully uncommon. Be mindful of the main sources: wild birds, rodents and even insects can pass disease to your chickens, but the most common culprit, by far, is the human. Shoes, clothing, tyres, unsterilised equipment and even unwashed hands are all perfect for spreading disease from one place to another, so adhere to basic hygiene at all times. Be especially aware when taking birds to shows.

Most of the time your chickens will be lively and content, but once in a while you may spot something different and, in the case of disease, noticing this may well save your chicken or indeed the whole flock. Look out for a drop in activity levels, eating or drinking less, diarrhoea, difficulty in breathing, sores, eye discharge, sore eyes, thin- or soft-shelled eggs, or birds standing with feathers puffed up and/or their head hanging or tucked under a wing. These signs may indicate the onset of a disease. That said, it is very likely that you will have a lifetime of chicken keeping with no occurrence of disease of any kind.

Coccidiosis This parasitic disease is caused by an intestinal protozoan, which feeds on chicks from the inside. It is most common in birds around a month old. Affected chicks go off their food, may have diarrhoea, become tired and shiver. Contact your vet immediately if you suspect coccidiosis as it is a speedy killer. It is easily treated with a drinking water medication; homeopathic remedies also work well.

Marek's disease A serious disease that usually results in death. Symptoms include paralysis of the legs or wings, tumours, blindness and general weakness. Vaccination is the best prevention and is commonly undertaken by suppliers of day-old chicks. There is no cure and even vaccinated birds can pass on the disease, so do choose your suppliers well.

Mycoplasma Also known as air sac disease, this bacterial infection shows itself in sneezing, gasping, running eyes and difficulty in breathing. Healthy birds are rarely troubled. It can be vaccinated against, and your vet can supply antibiotics to treat the disease if required.

Infectious bronchitis A respiratory disease that can quickly wipe out a flock, although it is not always fatal. Almost all chickens sold as day-old or point-of-lay have been vaccinated against this, but do check before handing over your money. It is one to vaccinate against if you intend to show, breed or sell your birds.

Gumboro disease This weakens the immune system, allowing other diseases to develop unopposed. It is not usually necessary to vaccinate against in small, closed flocks but definitely one to consider in other circumstances.

Newcastle disease An incurable viral disease that causes numerous problems, including serious breathing difficulties, tremors, paralysis and death in chickens, and flu-like symptoms and conjunctivitis in humans. If you suspect a chicken has this disease, even if it has died, you are legally required to report it to your vet and in turn to the appropriate authorities.

Salmonella This is a potentially serious bacterial infection that can affect humans. However, it is relatively uncommon and many chickens from large-scale suppliers (including ex-battery hens) will have been vaccinated against it. The most usual form of transference from chicken to human is via the droppings – usually from particles on the outside of the shell falling into a cracked egg. Common-sense precautions are the key: wash the eggs when you bring them into the house and wash your hands with hot water and soap after collecting and washing eggs.

Avian flu Also known as bird flu, this is surprisingly common: where there are wild birds there is avian flu. It comes in various strains, some of which cause death, but it is very unlikely that your chickens will contract it. Of particular concern is the H5N1 strain, as it is transferable to humans, potentially causing severe illness and even death. Thankfully, this strain is extremely rare and as yet its ability to transfer between humans is unsubstantiated.

Signs of avian flu include a swollen head and/or wattle and comb, unsteadiness and/or paralysis, unquenchable thirst, lack of appetite and difficulty in breathing. Contact your vet immediately if you suspect any of your chickens have avian flu.

Pests and predators

It is not only you who will view your chickens as providers of a tasty meal: foxes, badgers, birds of prey and even neighbouring dogs are among those happy to have a munch at your and the chicken's expense. Chicken feed makes good rodent food, as you'll soon find out if you store yours poorly or leave spillages uncleared.

Exclusion is your first priority when it comes to dealing with pests and predators. It is easy to get complacent about the threat, especially if you seldom see them. Think again. A fox is perfectly likely to come back night after night, ready to take advantage of the one time you don't lock your hens up or inadvertently let the battery run flat on the electric fence. Once in, they are likely to kill all the birds they can. What's more, they will return again and again. Imagine and act as if they are out there every night; they probably are.

A strong house is a must and any enclosure needs to be sturdy. Be aware of any potential overhangs that will allow a predator to jump in – a wall or trees give easy access to an enclosure. If you are using fencing, make sure it is electrified or well secured to the ground, or better still sunk into the ground to prevent burrowing.

Lock up your chickens promptly at dusk and if you lose any to aerial attacks from birds of prey, net over the top of the enclosure.

Also, consider the colour of the birds you choose if free-ranging them. We found our white Light Sussex birds were the first to be picked off from a large mixed flock, probably because their colouring made them so much more visible from a distance.

In extreme cases you may feel the need to have the predator trapped or killed. There is much sensible legislation around the use of guns and traps, so familiarise yourself with it before you consider this as an option. Trapping and/or shooting is fraught with complications, not least of which is who will (and is legally able to) dispatch whatever is tracked down or caught. Be aware that the likelihood is that other predators will take their place once the territory comes up for grabs.

It is far simpler to keep up your end of things and maintain sensible security measures than to enter into an ongoing war with predators.

Foxes I am very fond of the fox. It is an intelligent animal that, like you, has itself and very possibly a family to feed and it doesn't get distracted when it is carrying out its daily routine. It will not forget to visit your chickens when you fail to lock them up, neither will it overlook an opportunity to dig under fencing. Whether you are in the town or the country, assume foxes are nearby.

Most of their work is done at night, so prompt locking up of the henhouse, frequent visits and adequate fencing are all excellent deterrents. If you have a dog, take it for walks around your chicken enclosure – their urine will help mark the area as 'belonging to' another large animal.

Fox attacks are particularly likely in late spring and early summer when new litters may have been born and there may be little food around to feed them.

Badgers These strong animals are perfectly capable of breaking into most chicken houses to get at the birds. Attacks are fairly rare, but if you have them in your area, electric fencing is the best deterrent.

Electric fencing

Mink and weasels Both of these are comparatively uncommon these days. They usually live near rivers and streams and tend to kill by tearing off the chicken's head and drinking its blood. Attacks are rare as they are secretive creatures and the slightest noise or movement is usually enough to put them off. Fencing, a secure house and prompt locking up are the keys to minimising the potential for attack.

Other birds Birds of prey, seagulls and crows are very happy to dive in for a chick or egg for lunch, with larger hawks more than capable of taking an adult chicken. If you find this happening, consider enclosing your birds – especially the young ones – and fencing the top of their enclosure.

Discourage other birds from coming near your chickens by chasing them away. Some can transmit diseases to your birds, and they may eat the chickens' food.

Rats These can be a real problem. Look for large dark oval droppings, rat runs and/or holes chewed through housing and other structures. Rats will eat all the feed they can get at, urinating and spoiling the rest. They may well eat eggs and young chicks too, passing disease on as they do so. They can even chew into the structure of the housing and nibble through wires.

Traps can work to a degree, but you're then faced with the problem of dealing with trapped rats. Poison is very effective but it needs to be used safely; follow the instructions on the container. Use bait stations to protect other animals and humans from contact.

Mice Less harmful than rats, these rarely interact with chickens, other than to eat or spoil the food with their uncontrolled urinating. Their main calling card is their small oval droppings, but you may also notice their ball nests and small holes in your chicken feed bags. Poison is an option for eradicating them, and specialist traps that hold the bait inside are relatively safe from other animals and humans. If you don't want to kill mice, the traps can be baited with chocolate or peanut butter instead of poison, but do relocate the mice miles away otherwise they may well return.

Weevils, worms and moths A variety of these creatures are partial to chicken feed. Although largely harmless to the chickens themselves, they can cost you money and the birds valuable nutrients by stripping the feed of much that is beneficial to the chickens. You may notice these nuisances in the feed, or see fine webbing or tiny worms. Buying food only for the next 8 weeks ahead and storing it carefully is the best way to minimise the likelihood of trouble. If you discover a problem, discard the feed and scrub the feed bin, feeders and scoops with hot water and detergent, allowing them to dry thoroughly.

Unplanned deaths

Sooner or later one of your birds will die. If it is not at your own hand, it will almost certainly be due to natural causes brought on by old age. It may take you by surprise but in the weirdness of it all, think and act calmly. As a matter of routine, when a bird dies you should:

- Put on disposable gloves to prevent any possibility of disease being passed on to you.

- Move the corpse from the main enclosure and/or vicinity of the other birds straight away.

- Examine it in a well-lit area.

- Look for signs of an attack by a predator.

- Try to work out whether the bird's neck is broken – this may be caused by panicked action as much as physical attack by a predator.

- Identify any areas of missing feathers.

- Wash your hands thoroughly with soap and hot water.

- Dispose of the gloves.

Following the examination, check over the enclosure and any fencing to see if there are any places where damage has occurred or whether something has burrowed under the boundary and into the enclosure.

If you suspect disease is the culprit, act quickly. Your vet should be able to carry out (or at least refer you to someone who can carry out) an animal autopsy, known as a necropsy. This should identify the disease (if there is one) and make clear the course of action to follow.

In the unlikely event that your bird has died of a notifiable disease, you can expect an Official Veterinary Surgeon to visit and inspect your flock.

In any event, a dead bird should be disposed of through licensed premises, such as a local hunt kennel. DEFRA's Animal Health Department (see Directory, p.246) and/or your vet can advise as to your local options.

Dispatch

Every chicken owner should learn how to dispatch a bird.

You may want to see the whole journey through and dispatch your meat birds when they're ready for the table, or take care of a cockerel that's become a nuisance. The need to dispatch can come unexpectedly too: what happens on a wet Friday night when a bird is seriously ill or injured and the vet is unavailable? Also, if you are raising chickens from eggs or chicks, you should be prepared to dispatch very sick birds, any chicks born with an untreatable ailment and any unwanted cockerels. Even in a small flock of laying hens, illness and injury can happen at any time, and the need to dispatch a chicken can be urgent. Knowing what to do, even if you don't have plans to employ the skill, is part of what makes keeping chickens a more relaxed affair for you and provides a more humane end for them.

Having raised your own meat chickens, you may be determined to carry out every part of the process and dispatch your birds yourself, or you may be equally sure that you want to pass on that responsibility to a professional. Either is perfectly, equally fine. Please don't feel obligated to bring about their end, that it is in some way more morally right that you do so. Frankly, this isn't about you, it is about the chicken meeting a stress-free, quick end. If you don't want to do it, take the abattoir route.

If, however, you feel driven to be the person who does the deed, it is very much something you should feel able to do. It is straightforward and the task gets easier with repetition.

Preparing for a home end

If you are considering dispatching the chickens yourself, there are a number of things you need to bear in mind:

- You need outside space for the task itself – this is a messy business.

- Washing hands and tools as well as swilling out the kill area means you will need hot and cold water.

- You will have guts, feathers and blood to dispose of. Have a plan for their disposal before you decide to dispatch any bird(s). You can put these body parts into a pair of doubled-up bin bags but they should be disposed of pretty quickly, using someone recommended by your vet – a local hunt or abattoir may be able to help.

- Although dispatching is a quick job when you are experienced and confident, your first kill may take some time to gear up for and carry out. Allow at least an hour of unhurried time.

- You are legally bound to dispatch your birds on the property/land where they were raised: you are not permitted to transport them elsewhere if you are doing the killing.

- If you have neighbours, they may not be happy at witnessing the dispatch of your chickens. Choose a site that is not overlooked and/or erect a screen if need be.

- Make sure the place where you are dispatching your birds is out of sight and earshot of the rest of your birds.

Time to dispatch

Early morning is the best time to dispatch a bird: it will be sleepy and therefore easy to handle and less likely to be stressed. If you can do it before daylight proper, even better – it will be asleep and the task much simpler.

If you are only dispatching part of the flock, consider keeping those to be dispatched in separate housing the night before, as this reduces stress for the birds and for you. Although your birds should always have access to water, don't allow them to eat the evening before dispatch – their crop will then be empty in the morning, which makes dressing (see p.174) easier.

Dislocation: Simple hand method

Dispatch methods

There are endless methods for dispatching a bird, and while a blunderbuss will undoubtedly achieve the main goal, it is unlikely to deliver on the other two primary aims: a humane, stress-free end for the bird and to leave the carcass in good condition for eating. Although most of us have heard the phrase 'wringing a chicken's neck', I've never come across anyone who has dispatched a chicken by twisting its neck in the way one might wring out washing. The best methods are dislocation and decapitation.

Dislocation The aim of this method is to dislocate the neck by making a space between it and the head, thereby breaking the blood vessels that feed the brain. There are two good methods of dislocation: the most common hand method and the broom-handle method (see overleaf).

Dislocation by hand You may feel nervous when you first do this, so take time to calm yourself and remember to act without panic yet decisively – it is better to pull off the chicken's head by being too firm than not do it well enough. Here's what you need to do:

- With your palm upwards, using your weaker hand, take your chicken by the legs, allowing it to hang down (pic 1). The bird may be calm immediately, but if not it should certainly stop struggling after a minute or so.

- Turn the breast towards you. Place the first two fingers of your stronger hand in a V position, one either side of the bird's neck, with the bird's beak facing the ground and the back of your hand near the back of the chicken's neck (pic 2).

- Keeping hold of the legs, twist the knuckles of the neck hand around and simultaneously pull down firmly and sharply (pic 3).

- You should feel a space appear behind the chicken's skull so that your fingers can come together (pic 4).

Hang the bird by its feet afterwards, so that blood will collect in the head and neck, prior to removing the head.

As with decapitation, some movement (from a few small twitches to a minute of flapping) after the act is usual but be reassured that the bird is dead – it is just activity caused by the nervous system.

Dislocation: Broom-handle method

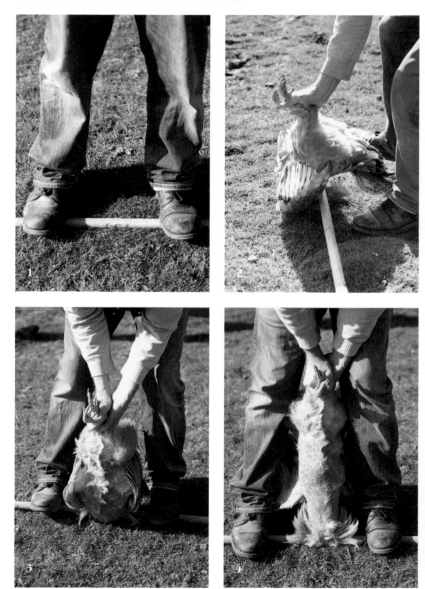

Dislocation by the broom-handle method This is surprisingly simple but easier with an assistant (as one of you can handle the chicken while the other manoeuvres the stick). You'll need a metre length of broomstick or similar.

- Place the stick on solid, level ground and position the balls of your feet on the stick about 30cm apart: this is the 'finish' position (pic 1).

- Holding the sleepy chicken by the feet, place it on the ground, chin down; it should be calm and fairly still. Facing the bird, place the stick across the chicken's neck, right behind the skull, keeping the stick lightly in place with a foot to one side of the bird's neck. Gently pull the chicken's feet to keep a slight tension so that the bird's head stays still against the stick and the neck extends a little (pic 2).

- Tread firmly down with both feet on the stick, one foot on either side of the chicken's neck (pic 3).

- At the same time, pull the body of the bird upwards and towards you (pic 4), if possible holding the wings in to restrain any post-death flapping.

As for the hand method of dislocation, hang the bird by its feet afterwards to allow blood to collect in the head and neck, before removing the head.

Decapitation While perhaps a little gruesome, decapitation is a simple way of dispatching a chicken that, providing your blade is sharp, is quick and decisive. It is, however, fairly messy. You needn't actually chop the bird's head off, you need only cut through the major artery, although severing the head does rather confirm that the task has been successfully undertaken.

To decapitate the chicken, choose a clean, sturdy log or stump to chop against and make sure you have a meat cleaver or axe that is very sharp. Have a killing cone (see Directory, p.246 for suppliers) or a hook and bag ready to hang and bleed the bird from. Fix a wire or strap across the log – the chicken's head will be placed under this to secure it.

When you're ready, carefully pick up the chicken, hold it by the legs and let it hang down. It may well be calm immediately, and if not it should certainly stop flapping about after a minute or so. Put the chicken on the stump with its head under the strap or wire, pulling the legs to extend the neck a little. Keeping hold of the chicken's body with one hand, strike down firmly with the blade just below the chicken's head.

Be aware that after decapitation a bird may well run about without its head: it is dead and the movement a matter of nerve and muscular memory. Rather than let go of the dead bird, place it either in a killing cone, or suspend it, legs tied together, from a hook, to bleed out. Again, there may be movement but be assured that, headless, the bird is dead.

Bleeding the bird

Immediately after dispatching, you should let all the blood run out of the bird. This makes the job of dressing your chicken (see p.174) much more pleasant.

If you have killed the chicken by dislocating its neck, the space that was created between head and neck will fill with blood if you hang the bird by its feet. If you prefer, you can place it in a killing cone or hold it by the feet over a bucket and cut its throat, letting the blood drain out of it completely. Bleeding is easy for a bird that has been decapitated – let it hang down in a killing cone or similar, allowing the blood to flow out of it until it stops dripping. Bag and dispose of the blood.

Although you can learn how to dispatch a bird from these pages, you might like to consider attending a course. Being shown how to do it and undertaking the killing with expert help makes matters easier for you (see Directory, p.246).

An abattoir end

There are a number of advantages in letting the professionals at an abattoir take care of killing and dressing your birds. You simply deliver them to the abattoir at the given time and return to collect your killed, plucked and dressed birds, ready to eat. You are spared the preparation, mess, clearing up and the stress, and you still get to eat fantastic chicken that you've raised yourself.

There are many abattoirs around the country that deal with chickens; the trick is to find a local one that's happy to take small numbers. Ask friends who keep chickens or local egg, poultry and meat suppliers if they know anywhere, or seek advice from your local Environmental Health Office, as they oversee small local abattoirs. You can even try small poultry producers in the area; they may have their own small abattoir on site and be happy to take care of your birds for you.

The downside of using an abattoir is the cost. The process of killing, plucking and dressing is fairly involved and the few chickens you are likely to want to send to them means the cost may be fairly high. Check before you commit.

When considering an abattoir for your birds:

• Ask other chicken keepers – a recommendation or two is always reassuring.

• Visit the abattoir before you make a decision, to gain an impression of it.

- Ask what the options are regarding the less popular parts of the bird. Giblets, liver and heart are usually returned with the dressed bird, but you may want the feet etc.

- Ensure you understand the procedures and timings for dropping your birds off and collecting them.

- Ask about packaging options – you may have to supply bags or they may provide their own.

- Decide whether you want your birds back whole or, if there is the option, whether you would like them jointed.

- Obtain a clear idea of the cost of each option and of the overall price.

- Book well ahead – busy times, especially at Christmas, can mean waiting several weeks for the next available space.

When it comes to planning the day you take your birds to the abattoir, allow yourself plenty of time. Loading birds can take a few moments or much longer if things don't go smoothly – knowing you are not against the clock if things aren't going immediately to plan helps to keep stress levels low for you and your birds.

If you can, get an early-morning slot to drop your birds off at the abattoir: chickens are sleepy and easy to handle when it is dark, and even if the sun is up, taking them from the house and into their crate is much easier than trying to catch them during the day.

Ensure you have proper carrying crates, and enough of them, in which to transport your birds (see p.51).

Dispatching chicks

If you raise chickens from eggs or chicks, it won't be long before you'll have to dispatch a chick. It will happen, so be properly prepared before the need arises. Decide whether you intend to strangle it or if you'd prefer to use dispatching pliers that work by being closed firmly around the neck from behind the head. It is, at least, physically easy to strangle a chick: simply grip, pull and twist the head away. Dispatching a chick is no fun, but when the time comes it is better for you and the unfortunate chick if everything is ready and you act quickly and calmly.

Preparing a Chicken
for Cooking

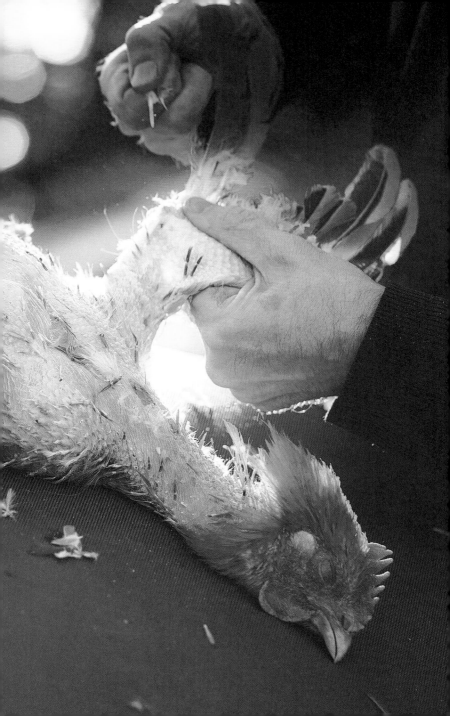

A dispatched chicken is still a long way short of oven-ready, and what you do after you kill it will have considerable influence over how good the bird is when you come to eat it.

Plucking your chicken

Allow time to remove the feathers from the chicken as quickly after dispatch as possible, certainly while it is still warm.

When plucking, start with the extremities as they usually cool quickest. Be aware that the skin can rip if you exert too much force, or more accurately, if you exert too much force in the wrong direction in the most sensitive areas. The areas to watch out for are the sides of the breast and the tops of the legs and wings – there are fat deposits here and the skin is most fragile. That's not to say there is anything wrong with tearing the skin here and there, but a bird for the table with an intact skin looks good. And, if you are thinking of selling your bird, rips just won't do.

There are dry and wet plucking methods. Whichever method you use may leave behind a few immature feathers, known as pinfeathers; these usually come out if pinched between fingernails and pulled. There may also be some fluffy, hair-like feathers remaining, which can be singed off with a lighter or similar. Dispose of the feathers, unless you're planning on making a small pillow any time soon – they will compost.

Dry plucking

This is the easiest and most sensible plucking method for the domestic chicken keeper. First, place the chicken on a suitable surface or sit down and lay the bird across your knees.

- Starting at the breast, take a couple of feathers at a time between thumb and fingers and pluck in the direction in which they grow (pic 1).

- Continue in this way to remove all the feathers from both breasts, taking care to pluck in the direction they lie (pic 2).

- Now pluck the feathers from the legs and wings. The wing feathers should be removed most carefully, one at a time (pic 3).

- Work along the back and up the neck, taking more feathers at a time and working against their grain until the bird is fully plucked (pic 4).

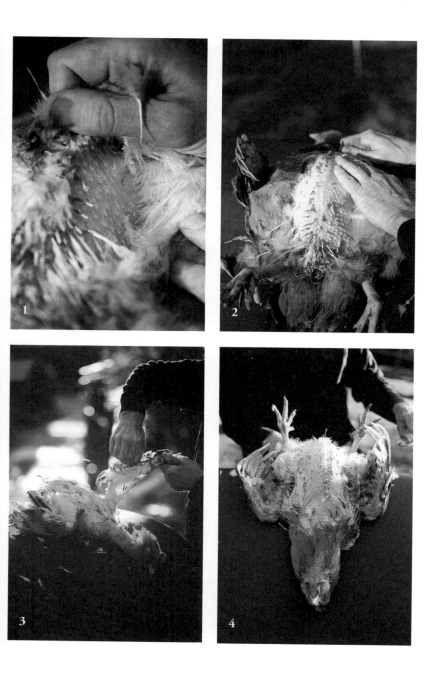

A freshly plucked chicken.

Wet plucking

Wet plucking can be quicker than dry but it does involve a fair bit more mess and means you'll have to dispose of a reasonable amount of unpleasant water. You'll need a container large enough to immerse your chicken in – a large sink or bath should work well – in an area that you are happy will get both wet and feathery. Rubber gloves can help with grip.

First, fill your container with water from a kettle, boiled and cooled to 55–60°C.

- Hold the bird by the feet and submerge all feathered areas for 15 seconds or so before pulling it out of the water.

- Starting with the wing feathers, pull in the direction of growth. They should come out easily; if not, submerge in the warm water for longer. Continue plucking until all the feathers are removed.

- Be prepared to change the water if it cools or gets too dirty and you need to re-submerge the bird.

Examining the plucked bird

At this point, it is worth looking over your chicken, partly to be sure it is healthy and also to become familiar with how a home-raised chicken looks and feels, which is often quite different from those you may be used to buying.

You may find the breast is smaller, the drumsticks darker and larger, the legs longer. Equally, they may not be. This is the start of getting to know your finished birds and how your management of them affects their final state.

Look over the skin. Depending on the breed, it should be somewhere between white and yellow. Don't be surprised if it seems a little loose: this is normal. The fat is often much yellower than on shop-bought chickens, owing to the varied free-range diet over a typically longer life.

You may notice the odd bruise – this can happen at dispatch. Any more than a few bruises is a sign that the chicken was being handled too roughly.

If you notice blisters on the breast skin, this usually indicates that their bedding wasn't changed regularly enough. Cut the blisters out, eat the bird as normal and change your birds' bedding more frequently.

Any other skin damage, such as an open wound or sores, indicates either a recent predator attack or disease. In either case, dispose of the bird.

Check the bird outside and in for abscesses. Whether they are caused by disease or wounds, these lumps of pus are not a good sign. They may indicate a wider issue if they are apparent in more than one bird – overstocking, perhaps. Dispose of any birds with abscesses – they are unfit to eat.

Dressing your chicken

The delightfully euphemistic term 'dressing' refers to pulling, cutting and by any other means relieving the chicken of the parts you'd better not or would rather not eat. These unwanted parts usually include the head, neck, feet and guts, but some of these can (and should) be saved for eating. The neck, gizzard, liver and heart are all undeniably delicious, and the feet, while perhaps an acquired taste for some, are enjoyed by many.

You can eviscerate your chicken immediately, as is the case commercially, or refrigerate your bird for dressing later. Much as it can be tempting to get the whole palaver out of the way quickly, I'd recommend the latter. A day in the fridge for your chicken has much to recommend it, firming up the flesh as well as the guts, while retarding the action of bacteria. Equally importantly, the process of hanging helps to tenderise the meat and allows the flavour to develop. It is a process that makes your chickens taste so much better than most of those you can buy.

To chill your bird before dressing, make space for it to hang by its feet if you can by removing some of the shelves within the fridge. Hang the bird upside down with the head contained in a plastic bag secured around its neck with a rubber band to catch any last drips of blood. If hanging isn't possible, lay the bird unwrapped and uncovered on a clean shelf with its head in the bag dangling off the end.

The bird should be kept in the fridge for 24 hours to chill thoroughly and it should be fine there for up to a week if your fridge has a fan to circulate the air.

When you are ready to dress your bird, take it from the fridge and discard the plastic bag. Have kitchen paper to hand to wipe up any blood, excrement, internal fluids – washing out with water isn't advisable as it can spread bacteria very effectively. Wash your hands before, frequently during and after dressing your bird.

Step by step guide to dressing a chicken:

- Feet first. Use secateurs to chop through below the drumstick, or use a sharp, heavy chopper. Alternatively slice around the joint and pull/twist the feet to tear them off, sinews and all.

- Remove the head if it is still attached, taking any neck skin that is damaged, but retaining as much as possible.

- Place the bird, breast down, on a chopping board and cut the skin from the top of the neck to the shoulder blades. Fold back the skin and remove the neck by pulling it up and cutting as close to the body as you can, twisting it free.

- With the neck skin folded back, look inside the bird to locate two tubes, the oesophagus and the windpipe. The windpipe is clean, stiff and easily removed by hand or knife (pic 1).

- Dealing with the oesophagus is more of a fiddle. Keep kitchen paper or some clean cloths to hand to deal with any digestive mess that may need clearing. Start by taking off all the small glands, then peel away the neck end of the oesophagus.

- Reach in with a finger and work around the oesophagus, separating it from connective tissues as thoroughly as you can, and discard it. Turn the bird over so it is breast side up, head end away from you, tucking the legs snug into the body.

- Expose the vent by pulling the skin and flesh upwards. With a sharp knife held parallel to the board, make a small slice across, through the skin only, just above the vent, to create access to the body cavity.

- Use your fingers to widen the hole and gently pull apart the skin and meat, until you find where the meat ends and the intestines and organs begin.

- Use kitchen paper to clean up any excrement that may come out.

- Use a small sharp knife to cut carefully around the vent, taking care not to puncture the intestines as you want to remove the two together.

- Pointing your fingers together, ease your whole hand in and under the breastbone, loosening all the tissues that connect the intestines as you go.

- Bend your wrist over a little, spread out your fingers and draw your hand back, bringing as much of what is loose as you can. Don't pull the entrails. They should come out in one (pic 2, p.175), although the lungs may remain attached at the top of the cavity and the kidneys towards the hole. If you can't get these out, either can stay in place as they don't affect the flavour of the bird.

- Tuck the legs and wings into the sides of the body, and secure the legs with string if you wish.

Your bird is now ready to prepare for cooking, freezing or sale. If you are eating it soon, keep it in the fridge until ready to cook. If you are freezing it but prefer to joint it first, see pp.177–9.

Wash and disinfect all surfaces and implements thoroughly when you have finished preparing the chicken and offal (see below).

Chicken offal Don't throw away the parts you have removed from the bird; therein lie some useful and delicious morsels, notably the heart, liver and gizzard.

The glossy, dark red/brown liver is easy to spot. Ease it away from the mass and cut away where you need to, taking care to avoid snagging the green gall bladder that is adjacent. Breaking the gall bladder releases a bitter liquid that can ruin the rest, so lose a little of the liver if you have to, rather than cut into the gall bladder. A trace of green on the liver, where the gall bladder rested, is quite normal. Check the usually red/brown liver for white spots or mottles – this indicates the likelihood of disease. Consult your vet if more than one bird has this and dispose of any affected bird.

The heart is also easy to recognise. Slip it out of the sac that holds it and slice off the line of fat. The gizzard is an island of kidney-shaped hardness in the mass of soft matter that remains. Slice off the in/out tubes and any fat. Cut around the edge of the gizzard, revealing a white line. Open it up and extract and discard the yellow sac that contains food remnants and stone used to grind it up.

Recipes for the offal can be found on pp.195, 228, 229 and 231.

Jointing and boning a chicken

Jointing a chicken is easy and liberating: it enables you to turn a whole chicken into smaller, manageable pieces for any number of recipes. It can also save you money, as it's much cheaper to buy a whole chicken and joint it than buy the constituent parts. If you raise chickens for meat, you may want to joint them before freezing, though I almost always freeze them whole and joint them after thawing if need be.

Give boning a try too (see pp.180–6). It is bizarrely satisfying to do and gives you a boneless carcass ready for stuffing, rolling, cooking and easy serving.

Jointing
Use a clean, sturdy chopping board and a sharp butcher's or boning knife. Follow the step-by-step photographic sequence on pp.178–9.

- Take hold of the wing and pull it out from the body in a circular motion to free it and make the join with the body more apparent. Cut through the skin and ligaments (pic 1). There's no need to cut through bones.

- Pull the leg slightly out from the body, and cut through the skin (and only the skin) along the line between the leg and the body (pic 2).

- Pull the leg further away from the body and downwards, pushing the ball at the base of the leg out of and slightly away from the socket (pic 3). Cut through the flesh that joins the body to the top of the thigh.

- If you want to separate the drumstick, bend the leg at the knee, cutting through where the crease forms (pic 4).

- Place the chicken on its back. Run your knife from the head end close along the breastbone to the tail end, feeling it clicking along the ribs (pic 5).

- Ease your knife under the breast and gently work it away, keeping the knife close to the carcass (pic 6).

- Release the breast (pic 7), then repeat to remove the other breast.

- This gives you 10 joints (pic 8) if the thighs and drumsticks are separated.

Pick off any meat attached to the carcass (freeze it if you haven't an immediate plan for it). The carcass is ready to use for making stock (see p.195).

Jointing a chicken

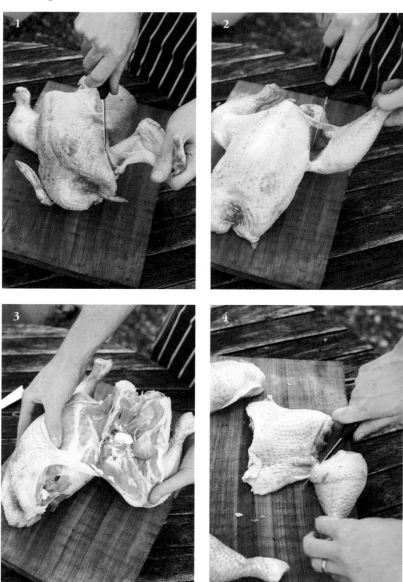

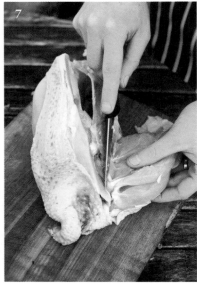

Boning

This may seem complicated but don't be put off: if you follow the stages slowly it is fairly straightforward. The secret is to use your knife sparingly – the technique is much more about using your fingers. Proceed as follows, taking your time and referring to the step-by-step photographic sequence on pp.182–5.

- Place the chicken, breast side up, on a board, legs towards you (pic 1).

- Pull out a wing and cut through the joint that is one 'knuckle' away from the body (pic 2). Repeat with the other wing. Set the wing pieces aside.

- Grip the neck skin and pull it back towards the breast – this will expose the flesh around the wishbone (pic 3).

- Make a cut either side, on the outside, of the wishbone. Get your fingers in there, feel the bone (which may be broken in two), and ease it out (pic 4). Be firm but not rough and mind the bone, as the end may be sharp. Pull the skin back down towards the neck.

- With the legs towards you, make a long cut through the skin, along the bone – this releases the skin on either side (pic 5).

- Focus on the left wing. Pull the skin of the wing back from the body towards the freshly cut wing 'knuckle' (pic 6). This should expose the flesh around the shoulder joint.

- The aim is to cut through the shoulder joint; wiggle the wing to clarify where it is. Make a cut through the point at which the two parts of the joint come together, wiggling the blade a little if you need to help it through between the bones and out the other side (pic 7). Repeat with the other wing.

- Sit the chicken up on its rump, back facing away from you. Take a wing firmly in one hand, and grip the body of the chicken with the other. Now pull the wing away from the body and downwards, making a circular movement as you pull, tearing the flesh from the carcass until you see the oyster exposed (pic 8). Repeat with the other wing.

- Turn the chicken around, so the back is towards you. Half of the skin and flesh should already be loose from the wings having been pulled wide. Place the first two fingers of your hand on either side of the central bone, knuckles pressed in towards the bird. Pull down, releasing the flesh from the carcass, until the whole front of the carcass is clear, apart from the fillets that cling to the central bone (pic 9); we'll take care of these in a moment.

- Lay your chicken down on the freshly cleared breast, neck away from you (pic 10).

- Hold the chicken by the flesh and skin around the leg, and cut around the oyster, adjacent to the backbone (pic 11). Bring the knee of the bird's leg upwards and with your thumb on the skin of the joint, take the leg back on the joint, towards you, to crack the joint and expose the bone. Cut through the sinew between the exposed bone and the carcass and pull the released bone and flesh back and away from the body. Repeat on the other leg. This should allow you to release the carcass.

- The carcass still has a fillet attached either side of the backbone and we want this lovely meat. Run a finger from the fat end of the fillet tight in against the bone to release the meat (pic 12). Repeat with the other fillet. Set the carcass aside for stock. Each fillet has a long white sinew running through it. To remove this, place the fillet with the sinew down, fat end towards you. The end of the sinew should be exposed at the fat end; if not, use the knife to reveal it. Holding the end of the sinew with a tea towel and the knife at a right angle to it, scrape away from you, against the board, to release the meat from the sinew. Repeat for the other fillet. Set aside the fillets; discard the sinews.

- Open out the bird, skin side down, legs towards you; there are the leg and wing bones still to remove. Take the end of the right leg bone where it is exposed (not the foot end) and cut around the knuckle just enough to be able to get a good grip on it. Hold the knuckle upwards and use the edge of the blade to scrape the flesh downwards and free from the bone, to the knee (pic 13). Use the blade to cut around the joint, and continue scraping until the shin bone is clear of flesh too but not cleared all the way to the foot end. The skin should still be attached to the foot end. Don't cut this or the skin will shrink up the flesh when it is cooked.

- Push the leg bones back into the leg, lean the shin bone up slightly, foot end against the board and use the back of a heavy knife to bash the bone about an inch or so up from the foot end (pic 14), breaking the bone in two. Pull the bone out, leaving the short foot end still attached to the skin. Repeat with the other leg. *(Continued on p.186)*

Boning a chicken

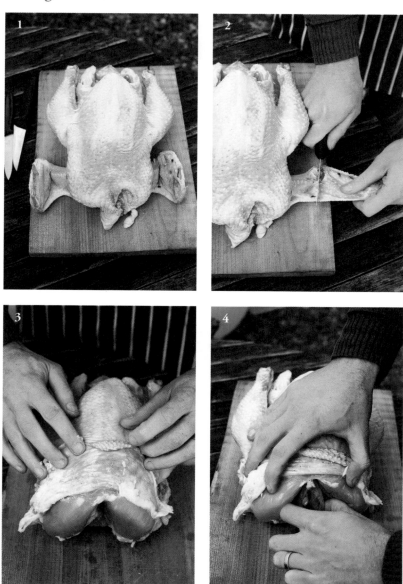

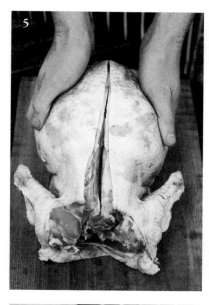

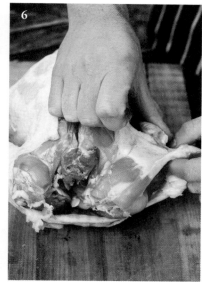

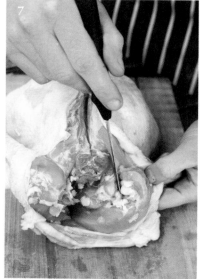

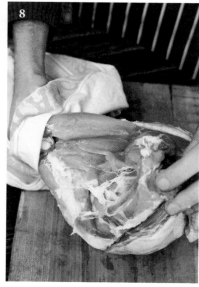

Boning a chicken (continued)

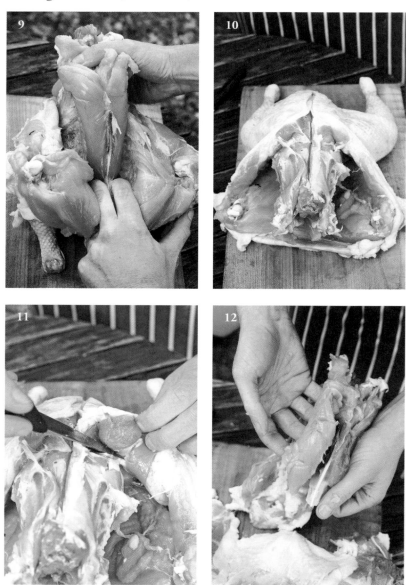

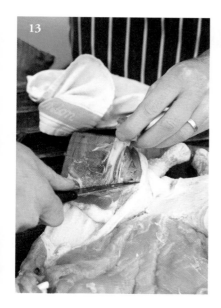

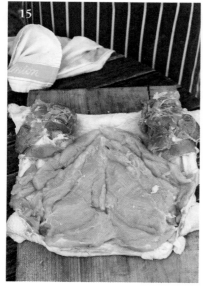

- Take a wing and, standing the exposed knuckle upwards, cut around the knuckle, then using your fingertips to close around the wing bone, press downwards, releasing the flesh from the bone. Pull the bone free. Repeat for the other wing. Ensure both wings and legs are turned so that they are not inside out. Lay the bird out open, wings wide and legs towards you. Ideally, you'll want a reasonably even covering of flesh, so lay the fillets in the fairly empty line on either side, between each leg and the breast (pic 15). If there's a gap at the neck end, you can always butterfly a little of the breast meat over to fill the space. Your chicken is now boned and ready for stuffing, grilling, roasting or cooking however you fancy.

- If you're stuffing the chicken (as for the recipe on p.208), once you've folded it around the stuffing, make sure you tie it evenly and neatly (pic 16).

Freezing chicken

Chicken, either whole or in portions, freezes well, although it begins to decline in quality after 6 months or so. Any in the freezer after a year should be discarded.

Tips for freezing chicken:

- Use good-quality, thick freezer bags that withstand low temperatures and protect the chicken well. Flimsy bags are likely to tear.

- Choose the right size for the job: large bags for whole chickens, much smaller ones for giblets etc.

- Label each bag, dating it and identifying the contents (giving the rough size if it's a whole chicken, or number of portions if jointed).

- Exclude as much air from the bag as possible as you seal it. Meat keeps best in the freezer with minimal air around it.

- Freezing chicken pieces or joints on a tray before bagging ensures they don't freeze together, and are more easily taken out to use a few at a time.

- Store all of your packs of raw chicken in the same drawer or basket of your freezer and rotate them (using the date information on each pack), so the oldest is always at the top and is the first one you reach for each time.

Storing eggs

Whether your eggs are for home consumption or for selling, you should develop a system for storing them and stick to it. Eggs will keep for around a month from the time of laying if they are kept cool, but some of the nutrition and flavour of your eggs is lost the older they get.

To keep eggs as fresh as you can for as long as possible, store eggs at around 5°C, ideally in a fridge, until a few hours before using, then take them out of the fridge so they can come to room temperature before you cook them. Keep them at cool room temperature if they are for your own consumption by all means, but they won't remain fresh for as long.

Create a rotation system that works for you. We keep a few egg boxes or trays handy and always add new eggs to the back of them, starting a new box or tray when it fills up and using eggs from the front of the foremost box. It means there is no danger of us using eggs out of synch with when they were laid. Write the laying date on the egg in pencil if you wish.

And remember: if you are not sure when any eggs were laid, keep them separate from the known eggs – if you've found them around the garden, you cannot be sure of when they were laid.

Freezing eggs

You can freeze eggs successfully, but not in their shells. Ice-cube trays with large compartments or very small containers are ideal. Separate the eggs into white and yolk: a yolk will usually fill one ice-cube compartment and the white two. Dividing the white can be a little fiddly, so fill the compartments of one tray with yolks first, collecting the whites in a bowl. Slide the yolk tray into a freezer bag to stop them drying out and place in the freezer, then put the whites into another tray. Or you may find it more convenient to freeze several egg whites together in the quantities you are likely to use for meringues etc.

Alternatively, you can lightly whisk whole eggs in pairs (or threes) and freeze in small tubs for using in cakes, omelettes, etc.

Label the containers, remembering to put the number of egg whites or whole eggs if you've packed them in multiples, and add the date they entered the freezer. You can expect them to remain edible for 6 months. Once defrosted, you can use them as you would fresh eggs but they may well lack a little of their unfrozen lifting power, so avoid making soufflés with defrosted eggs.

Pickling eggs

If you have an excess of eggs, try the recipe on p.227. Homemade pickled eggs are a real treat, altogether different from the ones you see in fish and chip shops.

Recipes

A meal is as pleasurable and nutritious as the ingredients
allow it to be. If you don't have chickens of your own, or you need to buy extra eggs or to supplement the meat you produce, it's good to know what you are getting.

Labelling on shop-bought chicken and eggs (or most produce for that matter) is, at best, unclear. It can be difficult to tell what you are paying for, and in turn what your money is going to fund. The following guide is intended to clarify the often bewildering choice.

Sourcing chicken

As with all meat, I would urge you to eat chicken of very good quality, even if that means eating it less frequently. The flavour and texture will be superb and it will be something you'll enjoy cooking and eating. Not only is it better for you than cheap and/or imported meat, it means your money makes the world of difference to the life of the chicken too. It may mean eating fewer chickens for your money, but the recipes on the following pages will help you get the full value from each bird – including making stock for risottos and soups, and ideas for using the leftover meat from a roast.

If you possibly can, buy organic free-range chicken. If not organic, then please make it free-range. One encounter with the realities of standard intensive systems is all it takes to put you off intensively reared chicken for life.

Free-range chickens have more room, freedom to roam outdoors and to express natural behaviours – in short, they are allowed to be chickens. And with many free-range chicken producers choosing slower-growing birds, it usually means the chickens enjoy a longer, as well as a happier, life.

Organic chickens are kept to free-range standards as a minimum requirement, but additionally enjoy organic GM-free feed and typically have a longer life.

If 'free-range', 'organic' or 'RSPCA Freedom Food' is not stated on the packaging, you should assume that the chicken has been raised in a standard intensive system. Phrases such as 'farm fresh', 'farm assured' and even the Red Tractor logo do not imply that the chicken has had a free-range life or indicate higher welfare. The extra money spent buying 'RSPCA Freedom Food' chicken ensures better conditions (including lower stocking rates) than most chickens enjoy, but the description can apply to indoor-reared chickens.

As ever, I would advise you to seek out a local respected producer or supplier. Your local butcher may be your best bet for buying chicken. Fine-quality butchers have a reputation to uphold that is easily ruined and most are aware that if they don't deliver on flavour and the welfare standards they advertise, customers have many options to turn to.

Sourcing eggs

Like chicken, eggs follow the organic, free-range, 'RSPCA Freedom Food' hierarchy, with all others indicating an unpleasant life for the laying hens in unsuitable conditions. Thankfully, labelling is clearer for eggs than it is for chickens: EU law ensures that it must clearly state how the hens that laid the eggs were kept. The use of phrases such as 'farm fresh' is still permitted but the eggs themselves will let you cut through the colourful language. Each individual egg must be stamped with an identifying code and the first digit in this code tells you what you need to know:

0 = Organic Birds laying these eggs have access to the outdoors and freedom to express natural behaviours, and they are fed an organic diet. No pesticides, herbicides or routine medicines are used.

1 = Free-range These chickens have access to outdoors and freedom to express natural behaviours.

2 = Barn-reared Birds that are kept indoors but not in cages, and have freedom to express natural behaviours.

3 = Caged Hens kept indoors in cages: add a postcard to an A4 sheet of paper and you have the minimum/typical floor area that each chicken has to itself.

As with chicken, look for a respected local supplier. Also, remember that many convenience foods are made with eggs: check the description and ingredients list on the packaging, and if it doesn't state specifically that it was made using organic or free-range eggs, it is almost certainly made with eggs from a caged hen.

Cooking with chicken and eggs

Having raised or sourced good-quality chicken and eggs, you'll want to do them justice in the kitchen. Most of us are familiar with how to roast a chicken and boil or fry an egg, but don't let this be the limit of your ambitions. Poaching a chicken gives you soft, succulent flesh and a lovely stock. Boning a chicken looks convoluted but it is pretty straightforward, and the results when stuffed and cooked are utterly delicious and an altogether different experience from a regular roast chicken.

Similarly with eggs, be prepared to poach, scramble and pickle them, and use them to make hollandaise, custard and meringues. Investigate more widely for yet more ways of enjoying the flesh and the eggs of the chicken.

Roast chicken

Even the smell of a chicken roasting in the oven can turn a wet, unwelcoming Sunday into a cosy restorative one. For a perfect roast chicken, there are a few rules: choose a good free-range, preferably organic, bird; don't overcook it; and let it rest for 15 minutes or so after roasting before carving or jointing.

Master the core recipe below and then embellish it as you like. A couple of my favourite variations are flavouring the bird with herb butter (see below), and roasting it with fennel, apples and bacon (see p.197), but more often than not I'll roast the chicken plain and simple. For a heavier bird, add an extra 3–4 minutes to the cooking time for every additional 100g.

Serves 4–5
1 chicken, about 1.6kg
2 tbsp olive oil
Sea salt and freshly ground black pepper

To serve
Chicken gravy (see p.194)

Take the chicken out of the fridge about an hour before cooking to bring it to room temperature, if necessary. Preheat the oven to 190°C/Gas mark 5.

Smear the skin of the chicken with olive oil and season generously with salt and pepper. Put the chicken in a roasting tray and place in the centre of the oven. Roast for 20 minutes, then lower the oven setting to 170°C/Gas mark 3 and roast for another 40 minutes. It should be nicely golden. To check that the bird is cooked, insert a skewer or knife into the thickest part, where the leg joins the body – the juices should run clear. If they are at all pink, roast the chicken for a further 10 minutes and check again.

When your chicken is cooked, turn off the heat, open the oven door and leave the bird to rest for 15 minutes. Joint or carve and serve with gravy and vegetables of your choice.

Variation

Roast chicken with herb butter Roughly chop a handful of thyme, tarragon or oregano, or the leaves from 3 rosemary stems with 6 sage leaves (or any mix of woody herbs you fancy). Crush or finely chop 4 peeled garlic cloves and mix with 80g soft butter (at room temperature) and salt and pepper. Work your fingers under the skin of the chicken from the neck end, freeing it from the flesh, then smear the herb butter all over the breast under the skin. Roast the chicken (as above).

Chicken gravy

To make the gravy, you will need nothing more than the pan in which your chicken has just been roasted, a glug of wine, a litre of stock (chicken or vegetable), and a little plain flour.

Serves 4–5

The residue in the roasting tray from
    a just-roasted chicken
A splosh of wine or Madeira
2–5 tsp plain flour

1 litre chicken stock (see right),
    or vegetable stock
Sea salt and freshly ground
    black pepper

Spoon or slowly pour off most of the fat from the roasting tin; do this carefully and the fat will remain separated from the tasty juices beneath, which should stay in the tin.

Place the roasting tin on the hob, over a moderate heat. When the juices begin to bubble well, add a splosh of wine and use a wooden spoon to scrape up any lovely burnt-on residue. When all is incorporated, add 2–3 tsp of flour, stirring vigorously with a whisk, adding more if needed to form a smooth paste.

Add 50ml of the stock to loosen the paste, whisking as you do so, and cook, stirring, for a minute or so. Gradually add the remaining stock, stirring constantly and returning the gravy to a simmer. Let it bubble to reduce down to the required consistency and concentrate the flavour. Season with salt and pepper to taste before serving with your roast chicken.

Chicken stock

You really have to make chicken stock – it is pretty much free food, providing the base for soups, risottos, stews, gravies and many, many other delicious meals. The ingredients do not need to be followed precisely – an extra carrot, less celery, leek tops instead of onions, a handful of parsley stalks will all be fine.

Makes about 1 litre

1 raw or cooked chicken carcass, stripped of meat
Chicken giblets, excluding the liver (optional)
2 onions, peeled and roughly chopped
2 large carrots, peeled and sliced lengthways
3 celery stalks, roughly chopped
1 tsp black peppercorns
2 bay leaves
A small handful of thyme

Take the chicken carcass and press down on the ribcage to crack the bones – this releases flavour from the bones and makes it easier to fit into a pan. Place the carcass in a pan in which it fits fairly snugly and that is deep enough to hold sufficient water to cover the bones.

Add all the other ingredients to the pan and pour in just enough cold water to cover everything. Bring to a very gentle simmer, then put the lid on and allow to simmer very gently for around 4 hours. Strain through a colander, discarding the bones and vegetables, then pass the stock through a sieve into a bowl and allow to cool. Use as required or freeze until needed.

Roast chicken with fennel, apples and bacon

The combination of the sweet/sharp crunch of the apples, the salty bacon and the gentle aniseed of the fennel with the chicken is simply stupendous. Try with quince in place of the apples too.

Serves 4–5

1 chicken, about 1.6kg
3 medium fennel bulbs
6 garlic cloves, peeled and crushed
    or finely chopped
50g butter
Freshly grated nutmeg
A handful of lemon (or regular) thyme

2 tbsp olive oil
6 rashers of streaky bacon, derinded
6 crisp eating apples
100ml white wine
Sea salt and freshly ground
    black pepper

Take the chicken out of the fridge an hour or so before cooking to bring it to room temperature, if necessary. Preheat the oven to 190°C/Gas mark 5.

Trim and finely chop the feathery tops from the fennel. Set the fennel bulbs aside. Mix the chopped fennel leaves with the garlic, butter and a good grating of nutmeg. Work your fingers under the skin of the chicken from the neck end, freeing it from the flesh, then smear the fennel butter all over the breast under the skin.

Put the thyme sprigs inside the bird's cavity. Smear the skin over the breast with olive oil, then season generously with salt and pepper. Lay the bacon across the breast and place the chicken in a roasting tray. Place in the centre of the oven to roast for 20 minutes. Once the bird is in the oven, quarter the fennel bulbs and cook in a steamer for about 10 minutes until just tender. Cut each of the apples into 6 wedges, removing the pips.

After 20 minutes, take the chicken out of the oven and baste it. Scatter the fennel and apples around the bird and pour in the wine. Return to the oven, lower the oven setting to 170°C/Gas mark 3 and roast for another 40 minutes. The chicken should be nicely golden. To check that it is cooked, insert a skewer or knife into the thickest part, where the leg joins the body – the juices should run clear. If they are at all pink, roast the chicken for another 10 minutes and check again.

Once the chicken is cooked, turn off the heat, open the oven door and leave the bird to rest for 15 minutes. Joint or carve to serve.

Chicken basquaise

There are many versions of this fabulous Mediterranean dish of chicken, peppers, olives and rice. This one – with its paprika and chorizo – is happily influenced by the Basque country of northern Spain. It's a beautifully melting, hearty, yet not heavy one-pot supper that makes me very content.

Serves 4

1 chicken, about 1.6kg, jointed into
    8 pieces (see p.177)
25g butter
6 tbsp olive oil
2 red onions, peeled and sliced
3 large red peppers, quartered
    lengthways, cored and deseeded
130g cooking chorizo sausage, skinned
    and cut into 5mm slices
8 sun-dried tomatoes in oil, drained
6 garlic cloves, peeled and chopped

300g basmati rice
A good squeeze of tomato paste
½ tsp hot paprika
4 bay leaves
A handful of thyme, leaves chopped
350ml chicken stock (see p.195)
180ml dry white (or red) wine
2 lemons, cut into wedges
100g black olives
Sea salt and freshly ground
    black pepper

Preheat the oven to 180°C/Gas mark 4. Have the chicken joints ready to cook.

Heat the butter and 3 tbsp olive oil in a flameproof casserole or large frying pan. Brown the chicken pieces in batches on both sides, seasoning them with salt and pepper as you go. Don't crowd the pan – fry the chicken in small batches, removing the pieces to kitchen paper as they are done.

Add a little more olive oil to the casserole and fry the onions over a medium heat for 10 minutes, stirring frequently, until softened but not browned. Add the rest of the oil, then the peppers and cook for another 5 minutes.

Add the chorizo, sun-dried tomatoes and garlic and cook for 2–3 minutes. Add the rice, stirring to ensure it is well coated in the oil. Stir in the tomato paste, paprika, bay leaves and chopped thyme. Pour in the stock and wine. When the liquid starts to bubble, turn the heat down to a gentle simmer. Press the rice down into the liquid if it isn't already submerged and place the chicken on top. Add the lemon wedges and olives around the chicken.

Cover and cook in the oven for 50 minutes. The rice should be cooked but still have some bite, and the chicken should have juices that run clear when pierced in the thickest part with a knife. If not, cook for another 5 minutes and check again.

Coq au cidre

Traditionally coq au vin is made with a cock bird and red wine, and very fine it is too. This adaptation using cider is, I think, every bit its equal. It works perfectly well with any hen or cock, and it is even better the day after. You can cook it in the oven (at 160°C/Gas mark 3) once you've added the cider, if that's more convenient.

Serves 4

1 chicken, about 1.6kg, jointed into
    8 pieces (see p.177)
50g butter, softened
3–4 tbsp olive oil
150g pancetta or unsmoked streaky
    bacon, cut into small cubes
10 eschallots or large shallots, peeled
8 garlic cloves, chopped
A good handful of thyme

4 tbsp brandy (ideally apple brandy)
3 bay leaves
700ml dry cider
200g small dark-gilled mushrooms
25g plain flour
A handful of parsley, finely chopped
Sea salt and freshly ground
    black pepper

Have the chicken joints ready to cook. Heat half the butter and 3 tbsp olive oil in a frying pan and brown the chicken in batches on both sides, seasoning with salt and pepper; don't crowd the pan. Transfer all the chicken joints to a flameproof casserole that will accommodate them in a single layer.

Add the pancetta to the frying pan and fry until lightly browned, then remove with a slotted spoon and add to the chicken. Add a little more oil to the pan if it is dry and cook the shallots gently, stirring frequently, for 15 minutes until soft but not brown. Add the garlic and thyme, cook for 2–3 minutes, then add the brandy.

Tip the contents of the frying pan over the chicken in the casserole and add the bay leaves. Pour in the cider, cover and simmer gently for 45 minutes. Stir in the mushrooms and cook for another 15 minutes. Check that the chicken is tender and the juices run clear when the thickest part is pierced with a knife. If not, cook for another 10 minutes and check again. Transfer the chicken, bacon, onions and mushrooms to a warmed serving dish and cover with foil to keep warm.

Bring the cidery liquid to the boil and reduce it by about a third. Meanwhile, mix the flour and remaining softened butter to a paste. Add about half of it, in pieces, to the liquid, whisking all the time. Keep whisking the bubbling liquid to cook the flour and thicken the sauce, adding more of the paste if needed, to thicken it further. Pour the sauce over the chicken and serve sprinkled with chopped parsley.

Chicken with walnut sauce

A toasted walnut and egg sauce brings chicken and egg together to make a lovely partnership. Accompany with rice if you wish, although I usually go for a green salad or steamed sprouting broccoli to keep it a little lighter.

Serves 4

1 chicken, about 1.6kg, jointed into
    8 pieces (see p.177)
25g butter
5 tbsp olive oil
2 onions, peeled and finely sliced
300ml dry white wine
400ml chicken stock (see p.195)
2 tsp sugar
2 bay leaves
3cm cinnamon stick (or 1 tsp ground)

Freshly grated nutmeg
3 large eggs
120g walnut pieces
8 garlic cloves, peeled and chopped
A pinch of saffron strands
Juice of 2 limes
A handful of parsley (or a third mint,
    two-thirds parsley), finely chopped
Sea salt and freshly ground
    black pepper

Have the chicken joints ready to cook. Heat the butter and 3 tbsp olive oil in a flameproof casserole or large frying pan (large enough to later hold the chicken pieces in a single layer). Brown the chicken in batches on both sides, seasoning with salt and pepper as you go. Don't crowd the pan – fry the chicken in small batches, removing the pieces to kitchen paper as they are done.

Add another 1 tbsp oil to the pan and cook the onions over a medium heat, stirring frequently, until softened but not brown. Return the chicken pieces to the pan, in a single layer. Add the wine, stock, sugar, bay leaves, cinnamon and a generous grating of nutmeg. Bring to a simmer, cover and simmer gently for 25 minutes.

Meanwhile, boil the eggs for 10 minutes. Drain, cool under cold running water, then peel. Cut around the centre of the egg and separate the yolks from the whites.

Lightly toast the walnuts in a dry frying pan over a medium-high heat, shaking the pan. Add the remaining olive oil, then the garlic, and cook for a minute.

Put the walnuts and garlic, egg yolks, and a few spoonfuls of the cooking liquid into a food processor and whiz to a smoothish paste. Stir this into the pan with the saffron and lime juice. Cook, uncovered, for 10 minutes or so, until the sauce has thickened. Check the chicken is cooked by piercing the thickest part with a knife to see if the juices run clear. If not, cook for another 5 minutes and check again. Finely chop the egg whites and sprinkle them with the herbs over the chicken.

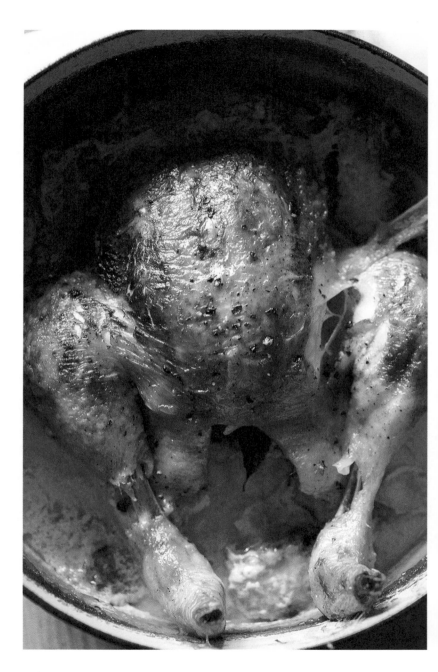

Lemon pot roast chicken

With three lemony flavours working together, this pot roast is something else. Like pork, poultry cooks beautifully in milk. If you're new to it, you might find it off-putting, anticipating something cloying, but it really is deliciously moist and melting. This recipe is also adaptable – try swapping the lemon thyme for regular thyme, add bay leaves, replace the lemon with cinnamon – but do try this version.

Serves 4

1 chicken, about 1.6kg
60g butter
2–3 tbsp olive oil
A handful of lemon thyme
About 12 sage leaves
8 lemon verbena leaves

10 garlic cloves, peeled and crushed
Finely pared zest of 2 lemons
  (in long strips)
About 600ml milk
Sea salt and freshly ground
  black pepper

Preheat the oven to 200°C/Gas mark 6.

Season the chicken generously with salt and pepper, inside and out. In a cooking pot that fits the bird fairly snugly, warm the butter with the olive oil until it begins to shimmer. Add the chicken and brown lightly all over, turning regularly. Take the pot off the heat and lift the chicken out onto a board. Tip out and discard the fat, but leave any caramelised loveliness on the base of the pan.

Put the lemon thyme sprigs inside the cavity of the chicken. Return the bird to the pot and scatter the sage, lemon verbena, garlic and lemon zest around it. Pour in enough milk to come halfway up the bird (you may need a bit more or less, depending on the size of your bird and pot).

Put the lid on and transfer to the oven. Cook for 20 minutes, then lower the oven setting to 180°C/Gas mark 4 and cook for a further 1–1¼ hours, basting the bird a few times. To check that it is cooked, insert a skewer or knife into the thickest part, where the leg joins the body – the juices should run clear. If not, return to the oven for a little longer. The result will be lemony, milky and succulent. The lemon zest causes the milk to split, giving a lovely yoghurty quality to the sauce.

To serve, pull the chicken flesh from the bones with a couple of forks – you won't need to carve. Sprouting broccoli is perfect with this, as are most greens. Spoon some of the lemony juice over the chicken and the greens.

Chicken with pomegranate
and perilla

Chicken pairs wonderfully with most fruit, sweet or sharp, including pomegranate seeds – a combination inspired by Diana Henry. Here these bright red seeds add punctuating crunch and flavour, while perilla's fabulous blend of mint and cumin flavours lends depth and freshness. If you can't find perilla (aka shiso), you'll find equal amounts of coriander and mint work equally, if differently, well.

Serves 4

2 tbsp olive oil
10g butter
8 chicken thighs (bone-in)
1 large onion, peeled and thinly sliced
1 tsp cumin seeds
½ tsp dried chilli flakes
¼ tsp hot paprika
8 garlic cloves, peeled and crushed

425ml chicken stock (see p.195)
150ml crème fraîche
½ pomegranate, seeds extracted
A handful of perilla (or coriander and mint), finely chopped
Sea salt and freshly ground black pepper

Heat the olive oil and butter in a wide frying pan. Season the chicken thighs with salt and pepper and brown on both sides, in 2 or 3 batches to avoid crowding the pan. Remove from the pan with a slotted spoon and drain on kitchen paper.

Add the onion to the pan and cook gently, stirring frequently, for 15 minutes or so until soft and translucent. In the meantime, toast the cumin seeds in a small dry frying pan over a medium-low heat until fragrant, then crush with a pestle and mortar. Add the cumin, chilli flakes, paprika and garlic to the onion and cook for a few more minutes.

Return the chicken thighs to the pan and pour in the stock. Bring to the boil, then immediately turn down the heat to a slow blip. Cover and cook for 20 minutes, then take off the lid and cook for a further 20 minutes. Test the chicken with a sharp knife – the juices should run clear. If not, cook for a little longer.

Stir the crème fraîche into the reduced cooking liquor. As soon as it begins to bubble again, remove from the heat, taste and adjust the seasoning. Scatter the pomegranate seeds and perilla over the chicken and serve.

Spicy stuffed chicken

A chicken relieved of all of its bones becomes a fabulous open envelope of meat to be filled with a tasty stuffing and tied. Easy to carve, moist and succulent.

Serves 6

1 chicken, about 1.8kg, boned
   (see p.180–6)
1 tsp cumin seeds
1 tsp coriander seeds
8 green cardamom pods
1 tbsp groundnut oil
30g butter
2 onions, peeled and finely diced
6 garlic cloves, peeled and finely
   chopped

3cm piece of ginger, peeled and grated
2 red chillies, deseeded and finely diced
2 tsp ground turmeric
1 tsp garam masala
A handful of coriander leaves,
   finely chopped
About 80g plain full-fat yoghurt
About 2 tbsp olive oil
Sea salt and freshly ground
   black pepper

Preheat the oven to 180°C/Gas mark 4. Lay the boned chicken skin side down on a board or clean surface and season with salt and pepper.

Toast the cumin and coriander seeds, and the cardamom pods in a large, dry frying pan over a medium-low heat for a minute or so to release their flavours and aroma, then pound with a pestle and mortar to crush. Pick out and discard the bits of pod.

Heat the groundnut oil and butter in a frying pan and fry the onions over a medium heat, stirring frequently, for about 10 minutes until soft but not browned. Add the garlic, ginger and chillies and cook for a further 5 minutes. Tip into a bowl, add the crushed spices, turmeric, garam masala and chopped coriander and stir well. Add just enough yoghurt to bind everything together to make a paste.

Spread the stuffing evenly over the chicken, then fold the sides over to meet in the middle and fold the loose skin at each end in a little to enclose the stuffing. Overlap the legs and tie them together with kitchen string, but don't cut it. Take the string under the bird, then over and under, pulling it through the loop. Repeat once or twice to tie hoops of string around the bird at intervals. Flip the bird over, take the string in a straight line under and round each hoop of string, then tie it at the feet.

Smear olive oil all over the skin, season well and place the chicken in a roasting tray. Roast in the oven for 1 hour. Test by inserting a knife or skewer into the middle of the meat. If the juices run clear it is ready; if not, cook for another 5 minutes and test again, until it is cooked. Leave to rest for 15 minutes before carving.

Chicken for the barbecue

Chicken soaks up flavours beautifully, especially with a little marinating, and is well suited to cooking over charcoal. Here are three of my favourite flavourings for barbecued chicken. They work equally well with chicken breast, drumsticks, thighs or chicken cut into pieces and threaded onto skewers for kebabs. I think skinless chicken works best, but the choice is yours.

Satay chicken

Serves 4
4 chicken breasts, or 8 drumsticks,
  thighs or wings, skinned
  if preferred

For the marinade
1 tbsp thin honey
1 tbsp soy sauce
A few drops of Tabasco
1 garlic clove, peeled and finely
  chopped
3cm piece of ginger, peeled and
  grated

For the satay sauce
1 tbsp groundnut oil
2 red chillies (medium heat), deseeded
  and chopped
3cm piece of ginger, peeled and grated
3 garlic cloves, peeled and finely
  chopped
100g crunchy peanut butter
2 tbsp soy sauce
Juice of 2 limes
1 tbsp light soft brown sugar
A handful of coriander, finely chopped
Sea salt

For the marinade, put all the ingredients into a bowl and whisk to combine. Add the chicken and rub the marinade into the flesh. Cover and refrigerate for a couple of hours to allow the flavours to permeate the chicken.

For the sauce, heat the groundnut oil in a frying pan over a medium-low heat, add the chillies, ginger and garlic and cook for a couple of minutes, then add the peanut butter, soy sauce, lime juice, sugar and a good pinch of salt. Stir well and add a little water to loosen the mixture if necessary – you're after a thickish, spreadable consistency. Stir in the chopped coriander. Taste and add more salt and/ or lime juice if necessary. Coat the chicken in the satay sauce.

Barbecue or grill the chicken, turning as necessary, until cooked through, about 15–20 minutes. Test by inserting a knife into the thickest part – the juices should run clear. Alternatively the chicken can be cooked in the oven, preheated to 180°C/ Gas mark 4, on a rack over a roasting tin to catch the sauce.

Jerk chicken

Chicken is fabulous flavoured with this aromatic, spicy Jamaican marinade for the barbecue or grill. Traditionally served with rice and peas.

Serves 4

4 chicken breasts, or 8 drumsticks, thighs or wings, skinned if preferred

For the marinade

1 tbsp allspice berries
1 tbsp black peppercorns
½ tsp ground cinnamon
½ tsp freshly grated nutmeg

A small bunch of thyme, leaves only, roughly chopped
6 spring onions, finely chopped
3 chillies (medium heat), finely chopped
1 tbsp dark brown sugar
1 tsp salt
2 tbsp dark soy sauce
Juice of 1 lime

For the marinade, pound the allspice and peppercorns together, with a pestle and mortar, then whiz in a food processor with the cinnamon, nutmeg, thyme, spring onions and chillies until smooth. Add the brown sugar, salt, soy sauce and lime juice and whiz briefly to combine.

Tip the marinade into a bowl, add the chicken and rub the mixture well into the flesh. Cover and leave to marinate in the fridge overnight, or for at least 4 hours.

Barbecue or grill the chicken, turning as necessary, until cooked through, about 15–20 minutes. Test by inserting a knife into the thickest part – the juices should run clear. Alternatively the chicken can be cooked in the oven, preheated to 180°C/ Gas mark 4, on a rack over a roasting tin to catch the juices.

Tandoori chicken

A wonderful way of cooking chicken in yoghurt flavoured with ginger, garlic and spices – aromatic and flavoursome without much chilli heat.

Serves 4

8 chicken drumsticks, thighs or wings, or 4 chicken breasts, skinned if preferred

For the marinade
Juice of 2 lemons
1 tsp cumin seeds
1 tsp coriander seeds
8cm cinnamon stick
8 cloves

3 garlic cloves, finely chopped
3cm piece of ginger, peeled and finely grated
1 tsp ground turmeric
1 tsp ground mace
1 tsp paprika
1 tsp cayenne pepper
400g plain full-fat yoghurt
Sea salt

Cut deep slashes in the chicken, but not through to the bone. Mix the lemon juice with a pinch of salt in a bowl, then add the chicken and rub the juice well into the flesh. Leave to stand for an hour or so, then drain the chicken.

Toast the cumin and coriander seeds, cinnamon and cloves in a frying pan over a medium heat, for about a minute to release the flavours and aromas. Grind the toasted spices, using a pestle and mortar, then add the garlic, ginger, turmeric, mace, paprika, cayenne pepper and 1 tsp salt and pound until well mixed.

Put the yoghurt into a large bowl, add the garlicky spice mix and whisk to combine. Add the chicken, rub the mixture well into the flesh, then cover and refrigerate overnight, or for at least 4 hours.

Barbecue or grill the chicken, turning as necessary, until cooked through, about 15–20 minutes. Test by inserting a knife into the thickest part – the juices should run clear. Alternatively the chicken can be cooked in the oven, preheated to 180°C/Gas mark 4, on a rack over a roasting tin to catch the juices.

Chicken curry

A lovely chicken curry that balances heat beautifully, letting all the flavours and aromas speak. It is, I'm immodest enough to suggest, a corker. You can, of course, poach chicken especially for the recipe, but one of its pleasures is that it gives you another meal from a chicken, without a great deal of effort. You can make the sauce ahead; do it in bulk and freeze some if you fancy. One last note: you can, of course, use all green cardamom pods, but you'll be surprised at the difference the black cardamoms make.

Serves 4

500g cooked boneless chicken
3 mace blades
9 green cardamom pods
6 black cardamom pods
1 tsp coriander seeds
4 tbsp vegetable oil
2 onions, peeled and finely sliced
3 bay leaves
10 cloves

½ tsp chilli powder
½ tsp ground turmeric
4 garlic cloves, peeled and finely
   chopped
3cm piece of ginger, peeled and grated
2 x 400g tins of chopped tomatoes
50ml double cream
A handful of coriander, chopped
Sea salt

Cut the cooked chicken into bite-sized chunks and set aside.

Gently toast the mace, cardamom pods and coriander seeds in a large, dry frying pan over a medium-low heat for about a minute to release their flavours and aromas. Tip them into a mortar and pound lightly with the pestle to break them up a bit.

Add the oil to the frying pan, increase the heat to medium and cook the onions gently, stirring frequently, until softened but not brown. Add the toasted spices, bay leaves, cloves, chilli powder, turmeric, garlic and ginger and cook for about 5 minutes, stirring occasionally.

Add the tinned tomatoes to the pan with a good pinch of salt, stir well and cook over a medium-low heat for 10 minutes or so. Remove from the heat and let cool slightly for a few minutes, then tip the contents of the pan into a food processor and whiz until smooth.

Return the sauce to the pan, stir in the cream, then add the chicken and bring to a very gentle simmer. Warm through for about 15 minutes and check the seasoning. Serve, scattered with coriander, with basmati rice and/or naan bread.

Candida's chicken pie
with chard and tarragon

This is another great way to stretch out the leftover chicken for a midweek supper, and the tarragon pairs beautifully with both the chard and the chicken. Feel free to make your own pastry but good-quality shop-bought is fine to use and makes this a very easy supper. You can use any veg or mixture that takes your fancy – fennel is particularly good, as is leftover veg from the roast.

Serves 6

300g leftover roast chicken
Juices from the roasting tray (optional)
400g Swiss (or ruby) chard
25g butter
2 tsp vegetable bouillon powder
  (Marigold)
2 rounded tbsp plain flour

600ml whole milk, plus a little extra
  for brushing
1 or 2 sprigs of French tarragon
240g packet ready-made shortcrust
  or puff pastry
Sea salt and freshly ground
  black pepper

Preheat the oven to 180°C/Gas mark 4. Roughly chop the chicken into bite-sized pieces and set aside with any saved roasting juices.

Separate the chard stalks from the leaves and cut the stalks into 2–3cm pieces. Blanch the leaves in a pan of boiling salted water for a minute or two, then remove with tongs and drain well. Add the chard stalks to the boiling water and blanch for 3–4 minutes until just tender, then drain well. Squeeze out excess water from the chard leaves and chop roughly.

Melt the butter in a saucepan, add the bouillon powder, stir in the flour and cook for a minute or two. Take off the heat and gradually add the milk, stirring all the time, to make a smooth sauce. Return to a low heat and slowly bring to the boil, stirring all the time. The sauce should be just thicker than double cream.

Add the chopped chicken to the sauce together with any roasting juices, the chard, tarragon and salt and pepper to taste. Heat gently for about 5 minutes, then tip the mixture into a 20cm round baking dish, 5–6cm deep.

Roll the pastry out thinly to a round or oval for the pie lid. Brush the rim of the pie dish with a little water, then place the pastry over the filling. Press the edge down onto the rim of the dish to seal and trim off the excess pastry. Brush the pastry lid with milk. Bake in the oven for 30 minutes until golden.

Poached chicken
and garden soup

Poaching a chicken is easy and provides you with a cooked bird and a delicious liquor that you can use for several meals – the cooked chicken can go to the curry (see p.215), the salads (on p.222 and p.224) or the noodle soup (on p.221). This fine soup makes excellent use of the liquor and as much of the meat as you wish to add. It's flexible and will welcome whatever seasonal veg are to hand.

Serves 4

1 chicken, about 1.6kg
A handful of flat-leaf parsley
3 bay leaves
A few sprigs of thyme
1 small celeriac, peeled and cut into
    5–8mm dice, leafy tops reserved
1 fennel bulb, trimmed and cut into
    5–8mm dice, leafy tops reserved

½ small Savoy cabbage, shredded
A good handful of peas (fresh or frozen)
Sea salt and freshly ground black
    pepper

To serve (optional)
2–3 tbsp crème fraîche
A little finely grated horseradish

Season the chicken with salt and pepper, inside and out, and put the parsley stalks (save the leaves), bay leaves and thyme sprigs into the cavity. Place the chicken in a cooking pot in which it fits quite snugly and pour on enough water to just cover. Slowly bring to the boil, then turn the heat down to a low simmer and put the lid on. Gently poach the chicken for about 20 minutes, skimming occasionally.

Add the celeriac and fennel and cook for a further 15 minutes, then add the cabbage and cook for another 15 minutes. To check that the chicken is cooked, insert a knife into the thickest part, where the leg joins the body. The juices should run clear and you should be able to pull the leg bone from the body easily. If not, poach for another 5 minutes and test again. When ready, add the peas and cook for a minute.

Lift the chicken out of the broth onto a board and leave until cool enough to handle. Meanwhile chop the parsley leaves with a little of the leafy tops from the fennel and celeriac. Tear as much meat from the carcass as you wish to add to the broth, leaving most of the poached meat for another meal.

Ladle the soup into warmed bowls, dividing the veg more or less evenly. Add the chicken and sprinkle with the parsley mix. Season with salt and pepper to taste. A dollop of crème fraîche spiked with a little horseradish will add a delicious bite.

Chicken noodle soup

This simple, aromatic, light, bright soup is one of the best ways of stretching a chicken out to make another dish and it is easily adapted to take in whatever vegetables you have to hand – broccoli instead of peas, a few shiitake mushrooms, shredded greens, etc.

Serves 4 as a starter

1 litre chicken stock (see p.195)
1 star anise
1 lemongrass stalk, bruised
150g dried udon noodles (or other kind if you prefer)
1 carrot, peeled and cut into matchsticks
A handful of peas (fresh or frozen)
200g cooked chicken, roughly shredded
1 tbsp olive oil
2 garlic cloves, peeled and finely chopped

3cm piece of ginger, peeled and finely chopped
3 spring onions, trimmed and finely sliced
Sea salt

To serve

A little spicy Tom Yum (Thai shrimp paste) or chilli sauce
A handful of coriander
A squeeze of lime juice

Pour the stock into a saucepan, add the star anise and lemongrass and slowly bring to a simmer.

Meanwhile, cook the noodles according to the packet instructions.

Fish out the star anise and lemongrass from the stock and discard, then add the carrot and peas. Bring back to a simmer and cook for a minute or so. Season with a good pinch of salt. Add the chicken and heat through for about 30 seconds.

Drain the noodles, rinse well with cold water, drain and toss with the olive oil. Divide between warmed bowls and scatter over the garlic, ginger and spring onions, dividing them equally between the bowls.

Taste the stock for seasoning and add a little more salt if required, then ladle the chicken and hot stock over the noodles and aromatics, making sure the chicken and vegetables are shared more or less evenly between the bowls.

Dab a few blobs of Tom Yum or a few drops of chilli sauce on each serving, scatter over plenty of coriander and add a decent squeeze of lime juice. Serve at once.

Chicken with rocket
and honey mustard dressing

A simple salad of contrasting flavours that makes a great quick lunch or, with a little more chicken, a sustaining supper. The amount of chicken is very much up to you – just a few scant shredded pieces, almost as a garnish, or more if you're in the mood. A few rocket flowers, if you have them, add colour and peppery punch.

Serves 2 for lunch

100g rocket leaves

½ red onion, peeled and very finely sliced

A handful of walnuts, toasted and roughly chopped

100–250g leftover roast or poached chicken (as you prefer), roughly shredded

Sea salt and freshly ground black pepper

For the honey and mustard dressing

2 tsp clear honey

1 tsp Dijon mustard

3 tbsp thick plain full-fat yoghurt

2–3 tbsp olive oil

To finish

A small handful of mint and coriander or perilla leaves

2 tsp perilla flowers (optional)

For the dressing, in a small bowl, stir the honey and mustard together until well combined, then add the yoghurt and mix thoroughly. Gradually stir in the olive oil to make a fairly thick, creamy dressing.

Pour two-thirds of the dressing into a large bowl, add the rocket and toss to coat the leaves in the dressing.

Lift a handful of rocket onto each plate, sprinkle with a little red onion and a few chopped walnuts, and add as much or as little chicken as you fancy.

Sprinkle with a pinch or two of salt and pepper, trickle over the rest of the dressing and scatter over the herbs and perilla flowers, if using.

St Clement's chicken salad

River Cottage head chef and all-round fine chap, Gill Meller, very kindly let me include his lemon-heavy courgette and sugar snap recipe in my *River Cottage Veg Patch Handbook*, and I have taken the spirit of it into this gorgeous salad. Here, the softness of the orange balances the sharp tang of the lemon without taking away its oomph.

Serves 4 as a starter, or 2 for lunch

100–250g leftover roast or poached chicken (as you prefer), shredded
4–6 baby courgettes, sliced into 3mm rounds
10 radishes, finely sliced
130g young, tender sugar snap peas
½ unwaxed lemon (cut vertically, not across)
½ orange (cut vertically, not across)
Sea salt and freshly ground black pepper

For the dressing
½ quantity mustard and honey dressing (see p.222)
OR
A little lemon juice and 2–3 tbsp olive oil

To finish
A small bunch of mint, tough stems removed, finely chopped
2 tsp perilla flowers (optional)

Put the shredded chicken into a salad bowl and add the courgette and radish slices with the sugar snaps.

Peel the lemon and orange halves, removing all the pith, then cut each segment free from its membrane, dropping the segments into the salad bowl. Don't worry if some of the segments break in half. Add the lemon and orange segments to the bowl, with any juice.

You can dress this salad however you like – it is very good with just a little squeeze of lemon juice and a splash of olive oil – but I think it's perhaps best with the honey mustard dressing.

Gently toss the salad in the bowl, then let it stand for 5 minutes to allow the flavours to mingle. Just before serving, toss lightly again, season with salt and pepper and scatter over the chopped mint and perilla flowers, if using.

Pickled eggs

Those Frankenstein eyes bobbing in a jar in the chippy are hardly the finest advertisement for the pickled egg (not that I mind them), but make them yourself and you'll see just how fine they can be. These are fantastic in a salade niçoise in place of the usual boiled egg, or as a picnicky snack, or dropped into a packet of crisps to pick up some of the salty seasoning. A great recipe to make when you have an excess of eggs.

Makes 8

8 eggs
125ml balsamic vinegar
600ml white wine vinegar
4 cloves
½ tsp coriander seeds
1 tsp Szechuan peppercorns
1 tsp black peppercorns
1 tsp black mustard seeds

A handful of thyme
2 large shallots, peeled and sliced
1 garlic clove, bashed
2 tbsp honey
3 sage leaves
1 bay leaf
1 tbsp salt

Have the eggs ready at room temperature.

Place all the ingredients except the eggs in a large saucepan, bring to the boil and simmer for 10 minutes. Take off the heat and allow to cool completely.

While the liquid is cooling, add the eggs to a pan of simmering water and cook for 10 minutes. Drain off the water and hold the pan of eggs under cold running water for a few minutes to halt the cooking process.

When the eggs are completely cool, peel them and place in a large, clean, sterilised jar (or two jars), but don't pack them in tightly. Pour in the pickling liquor to cover the eggs completely (if using two jars, divide the aromatics equally). Seal the jars.

Store the jars in a cool, dark cupboard to allow the pickled eggs to mature for at least a fortnight before eating.

Scrambled eggs
and chicken livers with sage

This is a great recipe for a quick and satisfying lunch. Try it first without toast – the combination of soft scrambled eggs and creamy liver may sound too lacking in crunch, but it is fabulous. Be careful to avoid overcooking the livers or eggs. The secret to perfect creamy scrambled eggs is to cook them very slowly, off the heat for part of the time, stirring as necessary.

Serves 2

400g chicken livers
4 tbsp olive oil
20g butter, plus an extra knob
1 onion, peeled and thinly sliced
6 sage leaves
8 eggs, lightly beaten

6 tbsp dry vermouth
A little Parmesan, freshly grated
   (optional)
A handful of chopped chives (optional)
Sea salt and freshly ground
   black pepper

Trim the chicken livers, removing any tubes and discoloured bits, and cut into roughly 5cm pieces.

Heat two-thirds of the olive oil with the 20g of butter in a frying pan, then add the onion. Fry over a medium-high heat to brown the onion slightly, without letting it get too soft.

Add the chicken livers to the pan, with the sage leaves, and cook the livers for 2 minutes on each side.

Lower the heat slightly and pour the beaten eggs into the pan. Stir, then as the eggs just start to cook, pour in the vermouth and turn off the heat. The eggs will cook enough in the residual heat of the pan; stir a few times as they cook.

Add the knob of butter and a little Parmesan and/or chives if you fancy. Season with salt and pepper to taste and serve straight away.

Chicken liver pâté

This easy pâté is delicious as a starter, lunch or snack. Beautifully savoury, very moreish and equally fine with good bread or crackers, it lasts no more than a few hours in our house, but it will keep in the fridge for 4–5 days and freezes well too.

Serves 6 (or more)

400g chicken livers
60g butter, softened
200g rindless streaky bacon, diced
2 large shallots, or 1 red onion, peeled and finely diced
4 garlic cloves, peeled and finely chopped
10 (or so) sage leaves, chopped

A few sprigs of thyme
Freshly grated nutmeg
3 tbsp Marsala (or port or sherry)
Sea salt and freshly ground black pepper

To seal (optional)
160g unsalted butter, melted

Trim the chicken livers, removing any tubes and discoloured bits, keeping them whole. Melt 30g of the butter in a wide frying pan until it just begins to bubble. Add the livers and fry quickly until coloured all over, but still lightly pink in the centre. Using a slotted spoon, lift the livers out onto a plate and set aside.

Add the diced bacon to the pan and fry until cooked but not crispy. Add a further 10g butter to the pan and gently cook the shallots over a low heat until soft and translucent – be patient, this can take 20 minutes or more.

Increase the heat a little and stir in the garlic, sage, thyme and a good grating of nutmeg. Now add the Marsala, tilt the pan a little and use a long match to set it alight. Once the flames have died down, add the remaining 20g butter. Allow it to melt, then remove from the heat and leave to cool for a few minutes.

Spoon the contents of the pan into a food processor and add the chicken livers. Blend briefly until combined but still fairly coarse-textured. Spoon into a bowl and season generously – the seasoning will be more mellow once the pâté is cool.

Line a small terrine or loaf tin or earthenware dish with a double layer of cling film, leaving some overhanging the sides. Spoon in the pâté and fold the overhanging cling film over the top. Refrigerate for 24 hours before eating.

After the pâté has cooled for an hour or so, you can uncover it and pour a thin layer of melted butter over the surface to seal and prevent it greying a little if you intend to keep it for a few days. Serve spread on hot, thin toast or thickly cut bread.

Chicken livers with Madeira
and caramelised onions

This fabulous quick lunch is infinitely adaptable. Try sautéeing 200g mushrooms along with the garlic and a little chopped sage. Or leave out the cream and toast and use the oniony livers as the base of a salad with rocket and sliced radishes.

Don't be tempted to overcook the chicken livers – 2 minutes either side is long enough to cook them while retaining their gorgeous creaminess.

Serves 4 as a starter, or 2 for lunch

400g chicken livers
50g butter
6 tbsp olive oil
2 red onions, peeled and thinly sliced
2 garlic cloves, peeled and finely
   chopped
8 tbsp Madeira (or port or dry sherry)

180ml double cream
Sea salt and freshly ground
   black pepper

To serve
4 generous slices of good bread
A handful of parsley, finely chopped

Trim the chicken livers, removing any tubes and discoloured bits, keeping them whole. Set aside.

Melt the butter with the olive oil in a frying pan over a medium heat. When hot, add the onions, season and cook for 5 minutes, stirring frequently. Turn the heat down a little and cook, stirring occasionally, for a further 10 minutes or so, until soft and caramelised. Stir in the garlic and cook for a further 3–4 minutes. Using a slotted spoon, remove the garlicky onions to a bowl.

Turn the heat under the pan up to medium-high. Add the chicken livers, season well with salt and pepper and cook, without stirring, for 2 minutes. Turn the livers and cook for a further 2 minutes – they should be brown on the outside and a little pink in the middle. Remove the livers with a slotted spoon and add to the onions.

Turn the heat up to high and pour in the Madeira. Set it alight with a long match to burn off the alcohol. As the flames die down, scrape up any burnt bits from the bottom of the pan and stir them into the liquid, using a wooden spoon. Add the cream to the pan and boil rapidly, stirring, for 2 minutes to reduce the liquid a little. Meanwhile, toast the slices of bread.

Pile the onions and liver onto the warm toast slices, spooning on as much sauce as you fancy. Sprinkle with chopped parsley and serve.

Spanish tortilla

When I worked in London, I was lured several times a day by the amazing aromas from a Spanish deli a hundred yards down the road. As often as not it was the onions softening and caramelising for their incredible tortilla and this is my attempt at recreating it. You need waxy potatoes that keep their shape when cooked. Strictly speaking, you don't need the chorizo – it is delicious without – but I do like to include it. Serve with a green salad for lunch. It is fabulous cold too.

Serves 4

125ml extra-virgin olive oil
400g waxy potatoes, cut into
    4mm slices
1 medium onion or 2 large shallots,
    peeled and finely sliced
60g chorizo, cut into 3mm slices

6 eggs
A small handful of flat-leaf parsley,
    finely chopped
Sea salt and freshly ground black
    pepper

Heat the olive oil in a 20cm non-stick frying pan over a moderate heat, add the potatoes and onion and fry for a few minutes. Put the lid on, leaving a little gap for steam to escape, and reduce the heat a little. Let the potatoes and onion cook gently for 25 minutes, stirring every 5 minutes or so. Add the chorizo and cook for a further 5 minutes.

Remove the pan from the heat and lift out the potatoes, onion and chorizo with a slotted spoon. Pour all but a couple of tablespoonfuls of the oil into a jug and return the potatoes, onion and chorizo to the pan, off the heat.

Lightly beat the eggs in a bowl, then pour over the potatoes. Add the parsley and plenty of salt and pepper, and stir well. Return the pan to the heat and turn the heat up slightly. Use a spatula to shape the tortilla, easing the top edge away from the side of the pan a little.

When the egg is almost set, take the pan off the heat. Invert a plate (one that is pretty much the same size as the tortilla) on top of it. Turn the pan and plate over together and lift the pan away, leaving the tortilla on the plate. Slide it off the plate back into the pan, cooked side up and let it rest off the heat for a few minutes to allow the base to cook a little in the residual heat of the pan.

Turn the tortilla out onto the plate again and allow to cool slightly and firm up for 10 minutes before serving.

Eggs Benedict

A fabulous brunch or breakfast – lighter than the full English, but every bit as satisfying. Traditionally this is made with ham, but smoked salmon is an excellent alternative if you fancy a change. Don't be put off by the hollandaise – it is simple to make and just incredible. When poaching eggs, many advocate adding a little vinegar or stirring up a vortex; I do neither, but don't let that stop you.

Serves 2

1 quantity of hollandaise sauce
   (see right)
2 muffins
4 large eggs

4 slices of Parma ham
A small handful of chives,
   finely chopped

Prepare the hollandaise sauce first and keep it warm in the bowl over a pan of hot water. Warm a couple of plates.

Two-thirds fill a wide medium saucepan with water, bring to the boil, then lower the heat to a very gentle simmer. Meanwhile, slice the muffins in half and toast them in your toaster or under the grill.

One at a time, break the eggs into a cup, then lower them carefully into the pan of simmering water. Let them cook in the gently bubbling water for 2½ minutes only (trust me).

While the eggs are cooking, spread a little hollandaise on each muffin half, then fold a piece of Parma ham on top.

As soon as the eggs are ready, remove each one with a slotted spoon, drain and dab underneath the spoon with a piece of kitchen paper to remove excess water.

Place an egg on top of each half-muffin and spoon a little more hollandaise over. Serve immediately, or, if you wish, give each plate a minute under a preheated grill to colour the hollandaise a little. Sprinkle with chives.

Hollandaise sauce

This is straightforward to prepare and not time-consuming. Once mastered, you'll find yourself looking for excuses to make it. Sharp and luscious, it is perfect with steamed asparagus or sprouting broccoli, and essential for Eggs Benedict (see left).

Serves 4

2 large egg yolks
1 tbsp lemon juice
1 tbsp white wine vinegar
A few black peppercorns

A sprig or two of tarragon
(optional)
120g unsalted butter
Sea salt and freshly ground
white pepper

Whisk the egg yolks in a bowl until they thicken just a little.

Put the lemon juice, wine vinegar and peppercorns into a small saucepan with the tarragon, if using. Let bubble to reduce for a couple of minutes until only about 2 tsp liquid remain, then strain. Very slowly pour the hot liquid onto the egg yolks, whisking as you do so to combine thoroughly.

Melt the butter in a small saucepan over a low heat until gently foaming but not colouring at all. Very slowly pour the butter onto the egg yolks, whisking constantly until it is all incorporated and the sauce has increased in volume. Add a good pinch of salt and pepper, scrape down the sides of the bowl and give it a last quick whisk. Check the seasoning and serve.

Custard

I am partial to the occasional idle dose of custard from a carton, or the nostalgia of custard made from powder, but nothing comes close to homemade custard. Allow yourself enough time to make it. If you don't want a skin on the custard (shame on you) then cover the surface closely with cling film until ready to serve.

Makes 600ml
600ml whole milk
1 vanilla pod
6 egg yolks
2 tbsp caster sugar
1 tbsp cornflour

Pour the milk into a heavy-based saucepan. Split the vanilla pod lengthways, release the tiny seeds with the tip of the knife and add them to the milk with the pod. Slowly bring to a bare simmer over a low heat and keep it at this temperature for a minute or two (don't let it even think about boiling).

Meanwhile, beat the egg yolks, sugar and cornflour together in a bowl. Lift out the vanilla pod from the hot milk, then pour the milk onto the yolky mixture, stirring vigorously as you do so.

Return the mixture to the rinsed-out pan and heat gently, stirring continuously with a wooden spoon, until the custard is thick enough to coat the back of the spoon. Don't be tempted to hurry along this part by upping the heat or not stirring – lumpy custard will be your reward. Pour the custard into a jug, ready to serve.

Meringues

Meringues are a great way to use up egg whites and they are very easy to make. They work equally well with fresh or frozen (and thawed) whites, too. For the best meringues, warm the sugar in the oven first to give you a stiffer mixture. Once made, you have many fine puddings to use them in, though I rarely get past filling a large round with fruit and cream as a pavlova, or breaking them up into pieces and folding them through whipped double cream with soft fruit for an Eton mess.

Makes 12 (or a large meringue round)
300g caster sugar
1 slice of lemon
5 egg whites

Preheat the oven to 200°C/Gas mark 6 and line two baking trays with baking parchment. Spread the sugar evenly over one of the baking trays. Place in the oven and allow it to just begin to melt around the edge but not colour; this should take 6–8 minutes.

Wipe the inside of an electric mixer bowl or large mixing bowl and your whisk with the lemon slice to remove any hint of grease. Add the eggs whites to the bowl and, just as you remove the sugar from the oven, start whisking at high speed (using the mixer or a handheld electric whisk) until they form soft peaks.

Turn the oven to its lowest setting, probably 100°C/Gas mark ¼. With great care, tip the hot sugar in a steady stream onto the egg whites, whisking as you do so. Keep whisking until the meringue is glossy and holds its shape. Add any flavours you might fancy at this stage.

Spoon the meringue onto the other prepared baking tray in a dozen luscious dollops (or a single great wheel if you prefer), spacing individual meringues apart to allow them room to expand as they cook.

Place in the oven for a couple of hours. The meringue(s) are done when they are crisp on the outside and lift off the paper easily. Leave to cool completely and store in an airtight container if not using immediately.

Variations

When the meringue is holding its shape, fold in 2 tbsp crushed, roasted hazelnuts or fruit purée (blackcurrant is especially good), or 1 tbsp sifted cocoa powder. Or, to give plain meringues a gentle honeycomb edge, use golden caster sugar.

Little orange soufflés

Although savoury soufflés are more common, I have to say I prefer their sweet sisters. These lovely orange soufflés have a dark golden, slightly chewy lid atop a soft, fluffy centre and a wonderfully foamy base. The recipe works perfectly with lemons too. I like to cook the soufflés in the hollowed-out oranges, as I find more oranginess is transferred to the soufflé this way, but the rise of the soufflé can occasionally be lopsided. If you prefer a more uniform rise, bake them in ramekins.

Serves 8 (or 4 if serving 2 each)

8 medium oranges
4 large eggs, separated
150g caster sugar

3 tbsp plain flour
A little icing sugar and/or cocoa
**   powder, for dusting (optional)**

Preheat the oven to 220°C/Gas mark 7. Line a baking sheet with baking parchment.

Cut the top third or so off each orange and a thin slice off the bottom so the orange sits flat on the baking sheet.

Scoop out the flesh from each orange, squeezing the juice into a bowl; set aside. Place the hollowed-out oranges on the baking sheet.

Put the egg whites and half of the sugar in the bowl of an electric mixer or a large mixing bowl and beat on a low speed (using the mixer or a handheld electric whisk) until the mixture becomes frothy. Gradually increase the speed, whisking until the meringue is glossy and holding soft peaks. It should only take a couple of minutes to get to the soft-peak stage; take care to avoid over-beating.

Combine the egg yolks, the remaining 75g sugar, 60ml orange juice and the flour in another large bowl and whisk until pale yellow. Whisk in a third of the meringue to loosen the mixture, then gently fold in the rest of the meringue, using a large metal spoon.

Spoon the mixture into the orange shells to just below the rim. Place the baking sheet in the top of the oven and lower the oven setting to 190°C/Gas mark 5. Bake for 8 minutes, until the soufflés have risen above the rims and the tops are golden. Don't open the oven any sooner to check them, as the cool air rushing in may cause the soufflés to collapse. If they need a little longer, shut the door quickly and give them another minute or two.

Remove the soufflés from the oven, sift over a little icing sugar and/or a little cocoa if you fancy and serve immediately – soufflés are just desperate to deflate.

Baked Alaska

This is a truly great, much under-appreciated pudding. Not only delicious, it has the added impossibility of ice cream in the oven. It is as simple as a cake or biscuity base, piled high with vanilla ice cream and smothered in a mountain of meringue.

Serves 6
For the base:
200g digestive biscuits
2 tbsp cocoa powder
100g butter

For the meringue:
300g caster sugar
1 slice of lemon
5 egg whites

For the ice cream:
1 quantity custard (see p.236)
or 500ml tub of your favourite vanilla ice cream

To make the base, break up the biscuits and blitz briefly in a food processor to coarse crumbs. Alternatively, place them in a strong plastic bag and pound with a rolling pin. Mix the biscuit crumbs and cocoa powder together in a bowl. Melt the butter in a pan over a low heat, add to the crumb mixture and mix to combine. Tip into a 20cm springform cake tin, spread evenly and press with the back of a spoon. Place in the fridge or freezer to chill.

For the ice cream, if you're making your own, churn the custard in an ice-cream maker until firm. Transfer to a tub and place in the freezer. Take the ice cream out of the freezer 10 minutes before assembling the pudding to let it soften a little.

If your oven has an integral grill, heat it to high. (Otherwise preheat the oven to 230°C/Gas mark 8.) Take the biscuit base out of the fridge/freezer and remove the side of the tin, leaving the biscuit on the base of the tin. Pile the ice cream on top of the biscuit base, leaving a 3cm clear margin around the edge. Put in the freezer.

Make the meringue (see p.238), whipping it to soft peaks. Take the ice-cream-topped base from the freezer and cover it evenly with meringue, creating a layer at least 5cm thick. This is vital: it insulates the ice cream, preventing it from melting.

Place in the middle of the oven and let the grill (or heat of the oven) colour and lightly cook the meringue. This can take anything from 3–6 minutes. After a few minutes, check frequently as it can quickly burn. When the contours are browned, with a hint of dark on the ridges, whip it out, immediately cut into it and serve.

Useful Things

Directory

If possible use recommended local breeders and suppliers. I have also found the following sources useful:

Suppliers

Harepathstead Poultry
www.harepathsteadpoultry.co.uk
Birds, housing and accessories.

P&T Poultry
www.pandtpoultry.co.uk
Housing, incubators, hatching eggs and accessories.

Brinsea Incubation Specialists
www.brinsea.co.uk
Incubators and brooders.

Perfect Poultry
www.perfectpoultry.co.uk
Eggs for hatching, chickens and equipment.

Omlet
www.omlet.co.uk
Eglu housing, runs, fencing, incubators, cleaning products, treatments and accessories.

Forsham Cottage Arks
www.forshamcottagearks.com
Hand-crafted wooden chicken houses.

BackYard Chickens
www.backyardchickens.com
An extensive resource, with a useful section of plans for self-build coops.

The Organic Feed Company
www.organicfeed.co.uk
Organic non-GM feed for layers, chicks and meat birds. Stockists in most areas.

Hi Peak Organic Feeds
www.hipeak.co.uk
Organic non-GM feed for layers, chicks and meat birds, delivered nationally.

Providence Farm
www.providencefarm.co.uk
Organic chicken and other meat, as well as courses (including dispatching birds).

Useful organisations

Chicken Out!
www.chickenout.tv
River Cottage and Compassion in World Farming, campaigning for higher welfare for chickens.

The Poultry Club of Great Britain
www.poultryclub.org
Charity that works to safeguard the interests of traditional chicken breeds.

DEFRA's Animal Health Department
www.defra.gov.uk/ahvla-en/
Advice for ensuring animal welfare.

British Hen Welfare Trust
www.bhwt.org.uk
Charity that re-homes commercial layers.

River Cottage Forum
www.rivercottage.net/forum/ask/poultry
Everything to do with poultry, user-run.

Acknowledgements

I wanted pigs; my wife wanted chickens. I think it's fair to say that we have inherited the other's enthusiasm. My daughter, Nell, took to both straight away. Thank you both for making it so much fun.

A decade ago, Pammy and Ritchie Riggs at Providence Farm took the time with a newly meat-eating me to talk about what they did, how their birds and pigs were raised and to answer any questions I had. They did much to inspire us to keep chickens. Their chicken, geese and ducks are still the benchmark by which we judge our own.

Thanks also to Paul Thompson at Harepathstead Poultry for his invaluable time and advice.

I am enormously grateful to Richard Atkinson, Natalie Hunt and Xa Shaw Stewart at Bloomsbury for their enthusiasm and dedication in making this book as good as it can be. Huge thanks also to Janet Illsley for her ever sensitive and insightful editing; to Will Webb, who designs this and the other handbooks so beautifully; and to Toby Atkins for his superb illustrations.

As ever, much gratitude to my agent, Caroline Michel at PFD, who appears to be made of energy and enthusiasm.

To Steven Lamb, thank you for the fun, the skills and the stunt hands. To Gill Meller, thank you for your fine touch and expertise on the photoshoots. To Rob Love, thank you… fancy some chickens to forage in that wild veg patch of yours? And, of course, huge thanks to Hugh, who has done so much to improve the chicken's lot.

Index

River Cottage Handbooks

Mushrooms

Preserves

Bread

Veg Patch

Edible Seashore

Sea Fishing

Hedgerow

Cakes

Fruit

Herbs

Chicken & Eggs

Seasonal, Local, Organic, Wild

FOR FURTHER INFORMATION AND
TO ORDER ONLINE, VISIT
RIVERCOTTAGE.NET